BIRD
PEN
MALAYSIA and
SINGAPORE

G.W.H. Davison and Chew Yen Fook

POCKET PHOTO GUIDE

BLOOMSBURY
LONDON · OXFORD · NEW YORK · NEW DELHI · SYDNEY

Bloomsbury Natural History
An imprint of Bloomsbury Publishing Plc

50 Bedford Square	1385 Broadway
London	New York
WC1B 3DP	NY 10018
UK	USA

www.bloomsbury.com

BLOOMSBURY and the Diana logo are trademarks of
Bloomsbury Publishing Plc

First published by New Holland UK Ltd, 2000 as *A Photographic Guide to Birds of Peninsular Malaysia and Singapore*
This edition first published by Bloomsbury, 2017

© Bloomsbury Publishing Plc, 2017
© text G.W.H. Davison
© photos Chew Yen Fook, except where otherwise indicated

G.W.H. Davison has asserted his right under the Copyright, Designs and Patents Act, 1988, to be identified as Author of this work.

All rights reserved. No part of this publication may be reproduced or transmitted in any form or by any means, electronic or mechanical, including photocopying, recording, or any information storage or retrieval system, without prior permission in writing from the publishers.

No responsibility for loss caused to any individual or organization acting on or refraining from action as a result of the material in this publication can be accepted by Bloomsbury or the author.

British Library Cataloguing-in-Publication Data
A catalogue record for this book is available from the British Library.

Library of Congress Cataloging-in-Publication data has been applied for.

ISBN: PB: 978-1-4729-3823-7
ePDF: 978-1-4729-3820-6
ePub: 978-1-4729-3822-0

2 4 6 8 10 9 7 5 3 1

Designed and typeset in UK by Susan McIntyre
Printed in China

MIX
Paper from responsible sources
FSC® C008047

To find out more about our authors and books visit www.bloomsbury.com.
Here you will find extracts, author interviews, details of forthcoming events and the option to sign up for our newsletters.

CONTENTS

Introduction	4
How to use this book	6
Glossary	6
Key to coloured tabs	7
Birds in the Malaysian and Singapore environment	8
Field equipment	11
Where to find birds	12
Species descriptions	14
Further reading	140
Index	141

INTRODUCTION

Birdwatching has come of age in Malaysia, just as it has in many countries around the world. What was once a hobby of the few has now become the basis of birdwatching groups in several local societies, which cater to their members with newsletters and field trips. Birdwatching has expanded away from a few key sites, to become more widespread. Some areas, such as Kuala Selangor in Peninsular Malaysia and Sungai Buloh in Singapore, are now protected and managed for the benefit of birds and other wildlife as a direct result of the initiatives of the birdwatching community. Bird races and other events are gaining popularity. Professional organisations dealing with nature and the environment are expanding.

In conjunction with this trend, there has been a considerable increase in the number of books dealing with birds in the region. The rising interest and the increasing availability of literature have been mutually supportive. This book adds to the list.

About 690 species of birds have been found in the area covered by the book, the political territories of Peninsular Malaysia (680 species) and Singapore (385 species). The political and geographical terms used to describe this area need some explanation.

The Malay Peninsula (a geographical term) is a long tongue of land appended to the south-eastern corner of mainland Asia. It extends 1,000km from the Isthmus of Kra to its southernmost tip. The upper portion of this peninsula belongs to Thailand, and its birdlife (while broadly similar to that further south) is not considered in this book.

The southern part of the peninsula belongs to Malaysia, and is commonly known as Peninsular Malaysia (a political term). It comprises 11 states and the capital, Kuala Lumpur. (The remaining two states of Malaysia, Sabah and Sarawak, are located on the island of Borneo, and are covered by an accompanying title in this series of books.) At the southern tip of Peninsular Malaysia lies the 622sq km island of Singapore, which is a separate nation.

This book describes and illustrates 252 of the 690 species found in Peninsular Malaysia and Singapore. They are generally speaking the commoner and more conspicuous birds to be found in each of a wide range of habitats, from mangroves along the coast to forest in the mountains. They also include, however, a variety of spectacular and interesting birds which have been covered because they characterise the bird fauna of the area and give a fuller flavour of the region's biological composition.

These 252 species therefore cover nearly all of the bird families or groups that can be found here. By familiarising yourself with the pictures, you should be able to identify a range of birds when you first encounter them. Be warned, however, that most of the groups include additional species not illustrated here. For example, there are six species of trogons in this area, of which two are illustrated; there are 20 species of pigeons and doves, of which 14 are illustrated. If you are to identify every bird, we recommend in particular Craig Robson's *A Field Guide to the Birds of South-East Asia* (2011, New Holland Publishers).

When identifying birds, it would be nice to think that comparison with a photograph is enough evidence to be sure of identification. In practice, however, a great deal depends also upon a bird's behaviour, its calls or song, and views at different angles or in different lights that reveal the full range of plumage features. Pictures alone cannot illustrate all of these features. It is therefore important, if you are to advance in birdwatching, to take notes on the birds you see and to make simple sketches of their features. From these notes and sketches it should be relatively easy to proceed to identify most birds.

Sometimes a single feature is enough to clinch identification, such as the curling red horn of the Rhinoceros Hornbill. More often, a combination of features is necessary in order to confirm a bird's identity, including size, general shape, shape and length of particular parts of the bird such as the tail, wings and bill, and colour pattern.

Much birdwatching can be done without any equipment. A notebook to jot down what you see is a basic requirement, and all except the beginning birdwatcher will require a pair of binoculars. Specialist advice on binoculars, telescopes and cameras can be found in a range of books and magazines. Most valuable of all, however, is advice from other birdwatchers. The beginner is likely to learn much more in a day spent birdwatching with an experienced friend than from a long time spent with books alone.

HOW TO USE THIS BOOK

The book has been designed with clarity and ease of use in mind. Opposite is a key to the symbols used on each page of the main species descriptions. These symbols are a guide to the family or group of families to which each bird belongs. Each such symbol first appears on the first full page bearing descriptions of that group. The photographs show the commonly seen plumage. Where there are two photographs, they are generally of male and female, with the male on the left or top. If they show other plumages, this is explained in the text.

The 252 species descriptions generally follow the same sequence as *A Complete Checklist of the Birds of the World* (see Further reading). There have been a few changes to names based on recent knowledge, and these largely follow Robson (2011).

The species descriptions begin with the common name, scientific name and length of the living bird from tip of bill to tip of tail. The next few sentences describe the main features of the bird needed for successful identification. This is followed by a sentence mentioning the calls and other behaviour, if distinctive and useful for identification. The next sentence begins 'Found in...' and describes the habitats in which the bird occurs. The final sentence begins 'Occurs in...' and gives the bird's distribution and its status in this region. A glossary of terms used in the descriptions is given below and the diagram on page 6 shows the names applied to the different parts of a bird.

GLOSSARY

Carpal joint The main bend of the wing, corresponding to the human wrist

Casque A helmet-like growth on the upper part of the bill, in hornbills and a few others, usually lightly constructed and hollow

Cere Bare skin round base of bill

Crown The top of the head, often distinctively coloured

Elfin forest Stunted montane forest

Flight feathers The long feathers of the wings and tail that form the main aerofoil surfaces during flight

Frugivore A fruit eater

Juvenile Young bird in its first full plumage, which often differs from that of adults

Mask Any distinctively coloured (usually dark) patch of feathers around the eye

Migrant A bird that undertakes long and regular journeys between its breeding and non-breeding areas

Moustache The area extending down and back across the face from the base of the bill, often distinctively coloured in groups of birds such as woodpeckers and sunbirds

Nape The point where the back of the head meets the back of the neck

Primaries The flight feathers on the outer joint of the wing, corresponding to the human hand, and in most birds forming the hindmost tip of the folded wing when the bird is perched

Rackets Expanded tips of feathers

Resident A bird that remains within a general area at all seasons, not undertaking long journeys

Scapulars Feathers over shoulderblades

Secondaries The flight feathers of the inner part of the wing, corresponding to the human forearm

Under tail-coverts Small feathers covering the bases of the tail feathers, beneath the tail

Upper tail-coverts Small feathers covering the bases of the tail feathers, above the tail

Wing-coverts Small feathers covering the bases of the primaries and secondaries on the wing

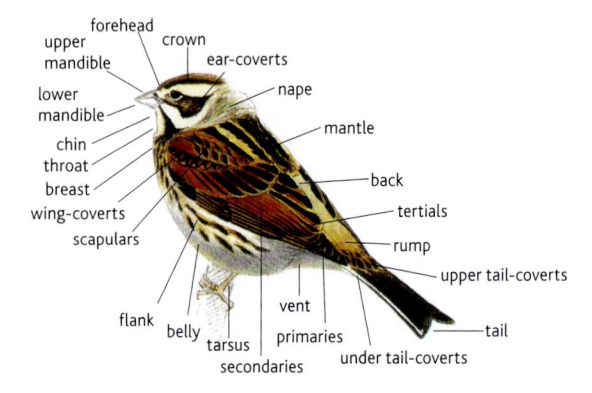

KEY TO COLOURED TABS

Grebes & boobies	 Herons & storks	 Raptors	 Gamebirds	 Rails & crakes, finfoot
 Waders & terns	 Pigeons & doves	 Parrots	 Cuckoos & relatives	 Owls
 Swifts	 Trogons	 Kingfishers & bee-eaters	 Hornbills	 Barbets
 Woodpeckers	 Broadbills	 Pittas	 Trillers & minivets	 Bulbuls
 Leafbirds	 Shrikes	 Wagtails & pipits	 Thrushes & relatives	 Babblers & relatives
 Warblers, prinias & tailor birds	 Flycatchers & relatives	 Nuthatches	 Flowerpeckers	 Sunbirds & spiderhunters
 Sparrows, munias & white-eyes	 Starlings, mynas & tits	 Orioles	 Drongos	 Crows

BIRDS IN THE MALAYSIAN AND SINGAPORE ENVIRONMENT

The Malay Peninsula (which is part of the Asian mainland), together with the great islands of Sumatra, Borneo and Java, are the main land masses on a fragment of the earth's surface known as the Sunda Shelf. Also included is the island of Palawan, which is politically part of the Philippines, and a host of smaller Malaysian and Indonesian islands.

The Sunda Shelf is an area of shallow seas, within which these main land masses are situated. The area has had a stable tropical climate for a very long time, being only slightly influenced by events such as the ice ages. At various times in the past, however, the Malay Peninsula, Sumatra, Borneo and Java have been linked by dry land, owing to temporary falls in sea level.

This means that there are many similarities but also some differences in the plant and animal life of the different territories. Also important in determining their composition is the physical geography of each region.

Peninsular Malaysia, occupying the southern two-thirds of the peninsula, has as its backbone the Main Range of mountains that reach a maximum of about 2,165m high. At various points along the range are the well-known birdwatching sites of Cameron Highlands, Fraser's Hill and Genting Highlands. To the west of the Main Range is a low coastal plain, some 50–80km wide. To the east is a broader lowland area, including low coastal plains and foothills, and Peninsular Malaysia's best-known freshwater lakes, a rare habitat in the region. Singapore, at the southern tip of the peninsula, is generally low-lying, with a central core of low hills.

Slope, altitude and drainage are some of the main factors influencing the natural habitats in this area, and thus their component species of wildlife. The map on p. 12 shows the approximate distribution of natural habitats in Peninsular Malaysia and Singapore. The altitudinal sequence of forest habitats, from mountain top to coast, follows the sequence from upper montane forest, lower montane forest, upper hill dipterocarp forest, hill dipterocarp forest, lowland dipterocarp forest, peatswamp and freshwater swamp forest, to mangrove forest on the coast. In practice, however, it is enough for most purposes to distinguish between montane forest above about 900m altitude (the first three of the detailed types just listed), lowland forest (the next four), and mangroves. These are three distinctive and easily recognised habitats, each of which has its own characteristic birdlife.

Lowland forest is still the most extensive of the forest types, but very little remains in the extreme level lowlands; most is now confined to hill slopes. What fragments remain of undisturbed extreme lowland forest are rich in wildlife and plants, and they are thought to be crucial to the survival of a core of about 40 bird species, such as Storm's Stork, Malayan Peacock-pheasant and Green Imperial-pigeon. Another 20 or so birds including as examples Great Argus, Green-billed Malkoha and White-crowned Hornbill, though they do occur in the extreme lowlands, prefer hill slopes within the lowland zone, that is up to about 900m. Fine birdwatching habitat in the extreme lowlands occurs in the national park Taman Negara, at Pasoh and in the Krau Game Reserve. Sites on hill slopes include Gombak and the Old Genting Road (see Where to find birds, p.11). In all about 240 species occur in lowland forest, of which between 180 and 210 are likely to occur at any one site depending on the forest's quality.

Montane forest is shorter in stature and poorer in tree species, cooler and damper with more epiphytes, orchids and mosses. About 71 bird species are largely or exclusively dependent on montane forest. All but one of them occur on the Main Range, but outlying peaks are impoverished; Gunung Tahan in Taman Negara has 50, as does Gunung Benom in the Krau Game Reserve, but Gunung Jerai on the west coast has only one truly montane bird species. The lower limit of the montane bird community is rather sharply defined, partly because there is no seasonal migration with altitude. However, the montane environment is enriched by the many lowland species which extend upwards into montane forest, especially in the absence of closely related competitors. For example, on peaks where the montane Blue Nuthatch is absent, the lowland Velvet-fronted Nuthatch may extend higher than it otherwise would.

Four birds are endemic to the Malay Peninsula. Three of these are montane: the Mountain Peacock-pheasant, Malayan Laughingthrush and Malayan Whistling-thrush *Myophoneus robinsoni*. The other, the Malayan Peacock-pheasant, is a lowland specialist.

In the mangroves, mainly on the west coast of the peninsula, there are up to 120 bird species. Of these, only about 25 are dependent on mangroves alone, the remainder occurring in other forested and non-forested lowland habitats. Some of the 25 depend on mangroves for nesting but feed elsewhere, and perhaps ten are absolutely confined to mangroves.

There is of course some movement between these habitats. A striking example is the Mountain Imperial-pigeon, nesting in montane forest but often flying to lowland forest or even the coast to feed by day, returning to the mountains to roost each night. Much less obvious is the vertical wandering within lowland habitats by a very few species such as the Little Spiderhunter, a bird which seeks nectar from scattered, sporadically occurring flowers such as wild banana. Where logging tracks or other forms of disturbance have entered forest, there may be intrusion by birds that otherwise inhabit open country, such as Oriental Magpie-robins. River banks are similar in this regard, and magpie-robins, Eurasian Tree-sparrows and some other birds associated with man have proven adept at colonising clearings or hill stations far from their usual range.

Food availability strongly influences movements within and between habitats. The abundance and accessibility of food in tropical forests show distinct seasonal fluctuations, even if not so extreme as in temperate countries. Some food sources, however, provide a tide-over for birds and other animals. Figs are a remarkable example: of 100 Peninsular Malaysian fig species, not all produce fruits attractive to birds, but those that do are excellent birdwatching hotspots. A variety of hornbills, barbets, pigeons, mynas and bulbuls, together with squirrels and primates, can usually be located at a heavily fruiting fig tree in the forest.

Another phenomenon associated with the tropics is the formation of mixed foraging flocks, in which a wave of birds passes through the forest. Perhaps a score of birds, of half-a-dozen species, associate in this way within a single flock. Some birds are consistently present in such flocks, others rarely join, and as a flock moves through the forest its membership will alter gradually as some individuals drop out and others join in. Flocks can occur at any time of year, but in this region seem to be commonest from about July to October. Encountering such a flock provides a rich haul of sightings for the birdwatcher.

Calls and songs are particularly important for birds in maintaining communication in the forest. Barbets and hornbills are amongst the first groups of birds that the birdwatcher will be able to distinguish in this way. Gradually the number of calls learnt will increase, but even an expert is unlikely to be able to name more than half of the calls heard in the forest.

Nesting seasons are well defined in this area, and fit around (although they are not necessarily determined by) other seasonal phenomena such as weather and food supply. In forest, peak breeding tends to be around March to May. The breeding season may be quite extended, with a long period of care of young after they have left the nest, but there are very few birds, if any, that nest all year round. Locating nests is difficult: the adults are secretive both when building and when rearing the young, and nests tend to be tiny, inconspicuous, and placed in precarious positions difficult for predators to reach.

In wetland habitats, herons and some other waterbirds begin nesting earlier than birds in forest, often about November or December, with young in the nest in the early months of the year. Other early nesters are the birds of prey, with the result that the slow-growing young may be in the nest when the supply of prey is at its most abundant.

Another seasonal phenomenon of great significance is the annual influx of migrants to forest (more than 30 species of flycatchers, cuckoos and others), to open country, and to the coast. More than 100,000 waders of 30 species use the west coast flyway, many of them on their way to Sumatra and further. Peak migration is in October and April, and a chain of mangrove and mudflat sites along the west coast from Kedah to Singapore is critical in maintaining this twice-yearly event. Kuala Selangor Nature Park in Peninsular Malaysia, and Sungai Buloh Nature Park in Singapore, have been developed for birdwatching and nature education as a response to this wonderful migration.

Although the broad outlines of birdlife and its biology are thus known, a great deal still needs to be done in recording, studying and conserving the region's birds. In this process birdwatchers will play an important and continuing role, for which identification is the first step.

FIELD EQUIPMENT

A notebook and pen should be considered basic and indispensable. Pencil smudges; ballpoint pens are good for permanency but do not work well on damp paper.

Rain and humidity are problems for binoculars; you can opt to buy an expensive brand, with waterproofing, or a cheaper brand for which fungus on the lenses is less heartbreaking. Binoculars of a specification around 8×30 or 10×40 are the most suitable. Wipe off damp binoculars gently after use, allow them to dry further indoors, and keep them ideally in a sealed container with silica gel.

Good equipment is bulky as well as expensive: specialist items such as cameras and tape recorders are not something beginners should consider until they are firmly committed to birdwatching.

Wear dull colours and lightweight clothing. You can either ignore rain, or use a poncho with a hood. Leeches will be encountered in most lowland forested areas, and can be ignored by the stout-hearted or picked off before they bite; insecticide sprays on footwear are effective but not good for the environment. Ticks can leave their jaws embedded, and carry disease (leeches do not); use fingernails (judiciously) or a canine shampoo to remove them.

WHERE TO FIND BIRDS

MONTANE FOREST

Cameron Highlands, Pahang. Access by bus or car through Tapah, Perak. Many places to stay. Extensive montane and hill resort with good birdwatching in forest, agriculture. Over 150 species.

Fraser's Hill, Pahang/Selangor. Access by bus or car via Kuala Kubu Baru or Raub. Several places to stay. Small hill resort with excellent birdwatching. Over 200 species.

Genting Highlands, Pahang/Selangor. Access by bus or car. Major hotels but often crowded. Some good birdwatching areas. Over 100 species. Also the Old Genting Road and Gunung Bunga Buah, Selangor (part of the Genting Highlands area). Access on foot. No facilities but excellent birdwatching along abandoned roadside and trails. Over 140 species.

Maxwell's Hill, Perak. Access by private landrover. Accommodation. Excellent birdwatching. Over 200 species.

LOWLAND FOREST

Kerau Game Reserve, Pahang. Permission needed from Department of Wildlife & National Parks. Limited facilities, but excellent virgin forest in extreme lowlands. Over 200 species.

Pasoh Forest Reserve, Negeri Sembilan. Permission needed from Forest Research Institute Malaysia. Limited facilities but excellent virgin forest in extreme lowlands. 200 species.

Taman Negara. The National Park. Good facilities and excellent virgin forest in extreme lowlands. Other forest types on longer trips. About 200 species in lowlands, over 270 species altogether.

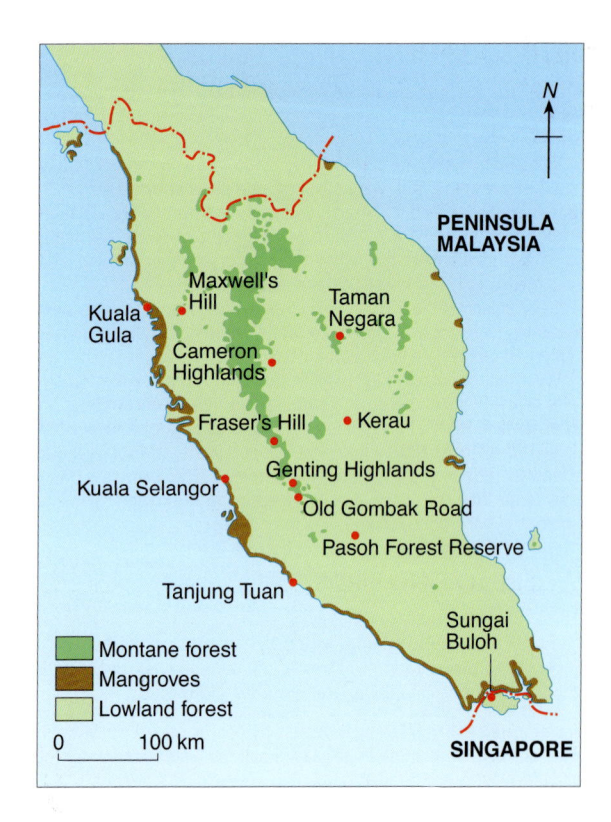

Old Gombak Road, Selangor. Disturbed forest along a little-used road. Good birdwatching, with over 150 species.

In addition, more than 70 sites have been designated as Recreational Forests for picnicking, most of them with forested streams, good for birdwatching away from areas frequented by the public.

COASTAL HABITATS

Kuala Selangor Nature Park, Selangor. Access by car or bus. With mangroves, brackish pools and secondary forest. Migrant waders in season and over 120 species altogether.

Sungai Buloh Nature Park, Singapore. Access by car. Very like Kuala Selangor in concept and birdlife.

Kuala Gula, Perak. With permission needed from Department of Wildlife & National Parks. Famous for Milky Storks and night-herons. Much other birdlife; over 120 species.

OPEN HABITATS

Try visiting the Lake Gardens in Kuala Lumpur, or the north and west coasts of Penang: abandoned mining pools and agricultural land including rice fields and oil-palm estates along the west coast; coastal scrub and coconut plantations along the east coast. For migrant raptors, Tanjung Tuan (Cape Rachado), on the Negeri Sernbilan/Melaka state border, is a coastal headland famed for Oriental Honey-buzzards *Pernis ptilorhynchus* in season, but tree-covered hills in the suburbs of Petaling Jaya and Shah Alam are also good.

Within the relatively small area of Peninsular Malaysia and Singapore there is not much variation in birdlife from north to south. However, Perlis in the extreme north has a drier, more seasonal climate than elsewhere, and some birds such as Lineated Barbet *Megalaima lineata*, Chestnut-headed Bee-eater *Merops leschenaulti* and Puff-throated Babbler *Pellorneum ruficeps* are a northern element that creep down into central Peninsular Malaysia. There is also a small southern element including some recent colonists, such as Savanna Nightjar *Caprimulgus affinis*, to be looked for in Singapore.

Acknowledgements

All the photographs in this book were taken by Chew Yen Fook with the exception of the following: G. Davison (28a, 114a), Oh Soon Hock (14a), M. Kavanagh (65b), Taej Mundkur (34b), K.W. Scriven (14b, 70a, 112b) and Dionysius Sharma (49c).

Photographs are designated a–d, reading from left to right and top to bottom.

The photographer wishes to give special thanks to Datuk Lamri Ali, Director, Sabah Parks, and Mr Francis Liew, the Deputy Director, for their generous accommodation during photography at Kinabalu Park, and Mr Alim Bium, the Park's Technical Assistant, for invaluable advice. We are greatly indebted to Malaysian Airlines, in particular Mrs Siew Yong Gnanalingam, Corporate Affairs Manager, for sponsorship of travel for Chew Yen Fook and Ken Scriven. We thank Innoprise Corporation Sdn. Bhd. and the Project Manager of Danum Valley Field Centre for permission to use their facilities and to photograph in the Danum Valley Conservation Area. We are extremely grateful to Ken Scriven for his invaluable advice and his wonderful companionship in the field. Dr David Wells provided much useful information. Chew Yen Fook gives many thanks to Taman Negara Resort and staff for their generous assistance in accommodation and advice, in particular Mazmadi Haji Mohamad, Roslan Abdul Rahman and Zamri Mat Amin; also to Ailynn Seah of SMI Hotels & Resorts for her ceaseless efforts in ensuring the best use of time in Taman Negara, Liau Beng Chye for his constant inspiration and encouragement, and Mr T. Guna of Maxwell Hill. Last but not least, he expresses profound gratitude to his wife, Siang, for her unwavering support in his work.

LITTLE GREBE *Tachybaptus ruficollis* 25cm

An active diver, this small duck-shaped bird pursues aquatic beetles underwater and can even fly after them when they break the surface. Small, dark grey-brown above and off-white below. When breeding, face and neck deep rich maroon, underparts dark, and yellow skin at angle of bill: the typical plumage seen in the region. Usually seen singly or in pairs, sometimes with young which both parents feed on insects, small fish and other aquatic life. Found on old mining pools, swamps, rarely on coast. Occurs throughout the Old World; here a resident possibly supplemented by some northern migrants.

BROWN BOOBY *Sula leucogaster* 72cm

When it dives headlong for fish, the cigar-shaped booby is transformed into a graceful lance. Adult dark brown with dark tail; white abdomen and undersurface of wings. Juvenile paler, with muddy brown mottling on abdomen; head and breast always darker than abdomen, and sharply defined. Nesting in colonies on a few small offshore islands, numbers varying widely and now only one site locally; feeding in surrounding seas, seldom inshore. Found on bare, rocky outcrops well out to sea. Occurs widely in tropical seas; resident on Pulau Perak.

GREY HERON *Ardea cinerea* 100cm

Although common elsewhere, it is worth recording anywhere in southeast Asia. Tall, bulkier than Purple Heron, grey above with black flight feathers and head patch, greyish white below with dark streaks down central neck and breast; yellowish bill and legs. Feeds typically on mud and at water's edge, waiting for and stalking fish, frogs, crabs. Often seen perched in trees or in slow flight. Forms nesting colonies in tall trees in and behind mangroves. Locally abundant in mangrove forest and adjacent coastal mudflats, seldom inland. Occurs almost throughout the Old World temperate and tropical zones; resident.

PURPLE HERON *Ardea purpurea* 95cm

Angular but elegant, this bird can be distinguished from Grey Heron, even in silhouette, by its slender appearance and snake-like neck. Slim, with head, neck and breast rufous cream, with black crown and stripes on face and neck. Back, wings and belly dark purplish-grey with some long rufous plumes. Typically solitary, quieter and less sociable than Grey Heron, seen perched on low dense trees and bushes, or motionless in shallows waiting for or stalking fish and other small animals. Scarce, forming small colonies in swamps and behind coastal mangroves, but feeding typically in fresh water. Occurs throughout the Old World tropics and warm temperate zones; resident and migrant.

GREAT EGRET *Ardea alba* 90cm

This is the largest of the egrets, most obviously so when in flight. Pure white, and with a strong bill. Non-breeding: bill yellow sometimes with dusky tip, face greenish-yellow, legs and toes black. Breeding: bill black, face bluish-green, thighs reddish or greenish contrasting with rest of black legs. Waits and stalks fish and other animals in shallow water. Found mainly on mangrove coasts, sometimes rice fields and swamps. Occurs commonly almost throughout the world; here a former scarce resident and now still a moderately common migrant.

INTERMEDIATE EGRET *Mesophoyx intermedia* 68cm

Less abundant than the Great and Little Egrets, this migrant is intermediate in size. All white, with plumes on breast and back (but not the nape) in breeding plumage, sometimes adopted before migrants depart. Short bill and bare facial skin always yellow; legs and toes all black. Bill length and softer curve to neck distinguish from Great Egret, yellow bill colour from Little Egret. Stalks prey on mud, in grass and shallow water. Found in mangroves, mudflats, rice fields and swamps, mainly coastal. Occurs from Africa through central and southern Asia to Japan and Australia; a migrant and winter visitor.

LITTLE EGRET *Egretta garzetta* 65cm

The commonest egret locally, and straightforward to identify. Pure white, with slender sharp-pointed black bill. Non-breeding: bill is virtually black, and legs black with yellow (rarely, also black) toes. Breeding: bill all black, face bluish-green; two long plumes on nape and filigree plumes on back and rump. Actively skitters while hunting fish and other small animals in shallow water. Found from coasts to inland freshwater of all types, but seldom in large aggregations. It occurs throughout the Old World; here it is most numerous non-breeding migrant heron.

EASTERN CATTLE EGRET *Bubulcus coromandus* 50cm

A recent taxonomic split from the Western Cattle Egret of Africa, Europe and the Americas. Small, thick-necked egret, pure white in non-breeding plumage with strong yellow bill and dark legs. Before breeding, head, neck, back and breast become suffused with buff, bill yellow, legs red. Often with cattle or buffaloes, feeding mainly on insects disturbed by them. Found especially near marshes, pools and rice fields, and grassland. Occurs widely from Pakistan in the west to Australia and Japan in the east; here a non-breeding migrant.

CHINESE POND-HERON *Ardeola bacchus* 45cm

In appearance half-way between a small heron and an egret, this bird is transformed on taking flight, from dark brown to pure white. Breeding plumage plain deep chestnut grading to black, with white wings, tail and abdomen. Non-breeding plumage, as typically seen in region, light brown above with head, neck and breast strongly streaked grey-brown and white; wings, tail and abdomen white. Seen singly, at water's edge in thick vegetation, hunting small fish and insects. Found in lowlands in freshwater swamps, rice fields, lakes, sometimes along coast. Occurs throughout east and south-east Asia; migrant.

LITTLE HERON *Butorides striata* 45cm

Both residents and, in the northern winter, an influx of migrants make up a big but dispersed population. Small, heavily plumaged heron, dark blue-grey or green-grey with nearly black crown; pale face markings, streaks on breast and narrow buff edges to wing feathers. Juveniles brown, stockier, streakier and less mottled than juvenile Cinnamon Bittern. Typically solitary, a stand-and-wait hunter, giving a single loud keyaw in flight. Found on streams, ponds, marshes, mangroves and seashore. Occurs in suitable habitat throughout the Old World tropics and subtropics; resident and migrant.

BLACK-CROWNED NIGHT-HERON
Nycticorax nycticorax 60cm

Active by day and night, this small stocky heron makes impressive dusk flights to its freshwater feeding sites. Three-coloured adult with black crown and back, pearly grey wings and white underparts. Juvenile (lower photo) dark brown, streaked buff below, and distinguished from young Cinnamon Bittern and Little Heron by bold buff spots on back and wings, by robust bill and stocky shape. It forms colonies, in mangroves; usually feeding a short way inland. Found in mangroves, mudflats, river edge and swamps. Occurs in almost all the world; resident.

YELLOW BITTERN *Ixobrychus sinensis* 37cm

Sometimes a burst of buff and black will erupt from between the reeds before dropping back into the swamp. The buff plumage with creamy buff wing coverts and black flight feathers give a bicoloured appearance in flight, and black cap. Juvenile has same wing pattern, body heavily streaked brown above and below. Skulking, seen singly in dense swamp vegetation, usually alone; feeding on small fish, frogs, invertebrates. Found in freshwater swamp, lotus, wet grassland, rice fields and old mining pools. Occurs throughout east Asia to New Guinea; resident.

MILKY STORK *Mycteria cinerea* 95cm

This rare and spectacular bird is in need of special conservation measures. Adult pure white with black flight feathers; heavy, slightly drooping yellow bill and deep red bare head and legs, greyer when not breeding. The juvenile is duller, milky brown with black flight feathers, with the body becoming gradually whiter with age. Now one breeding colony on west coast, where birds feed on mudflats, taking crabs, mudskippers and other animal life. Found in mangroves. Occurs from Peninsular Malaysia to Sulawesi (Celebes); here only resident at Kuala Gula, Perak, with occasional dispersants along west coast.

CINNAMON BITTERN *Ixobrychus cinnamomeus* 37cm

A slender little heron sometimes seen in flight across roads in swampy country. It is small, slim with a rich chestnut or cinnamon plumage, creamy beneath with dark streaks down centre neck and breast; yellowish bill and legs. In flight wings appear all dark, not strongly bicoloured. Juveniles darker brown, slim, with heavily streaked breast and mottled to even spotted wings. A strong flier but skulking on ground. Found singly but abundant in a wide variety of mainly freshwater wetlands, feeding on small fish and frogs. Occurs throughout east and south-east Asia up to 2,000m; resident.

STORM'S STORK *Ciconia stormi* 85cm

A Walt Disney bird, the adult has a striking clown-like red and yellow face. Confused identification with other storks in the region has only recently been clarified. Adult with red bill and legs, bare yellow face, and white plumage on upper neck and abdomen; remainder of head, neck, wings and body dark brown. Perches singly in riverside trees, occasionally descending to the swampy forest floor in search of frogs, or wheeling high overhead. Found in lowland forest, tree-lined swamp and riverside habitats. Occurs in south-east Asia from southernmost Thailand to Sumatra and Borneo; now a rare resident.

LESSER WHISTLING-DUCK *Dendrocygna javanica* 40cm

The commonest duck locally, a tree-nester whose young must fling themselves earthwards in their first few days of life. Rufous brown, rather long-necked duck, with pale buff cheeks, dark brown cap, and usually some pale streaking on flanks. They occur in small parties, diving for some vegetable and insect food, and taking floating weed from the surface. Pairs rush across water, erect with one wing raised, in display. Found on old mining pools, freshwater swamps, occasionally forested rivers and behind mangroves. Occurs in south-east Asia from India to Borneo; resident.

21

OSPREY *Pandion haliaetus* 54cm

A magnificent and evocative raptor, the Osprey is found nearly worldwide but is always a special sight, long, narrow wings giving unmistakable flight silhouette. Back and wings very dark brown, head and underparts white with dark markings or band across breast, and wide dark band through eye; dark tail faintly banded. Fish-eating, plucking live fish from the water and carrying them to exposed perch to eat; usually solitary, typically seen in flight or perched on fishing stakes or bare trees. Found mainly in coastal districts and on larger rivers and lakes. Occurs worldwide, at low altitudes; migrant.

BLACK-SHOULDERED KITE *Elanus caeruleus* 32cm

A most elegant and streamlined raptor, often seen from roadsides in rural areas. Adult pearly grey, almost white on the head and breast, darker on the wings and long square-ended tail; and a black shoulder patch. Juvenile duller, the grey back, wings and breast smudged with light brown. Seen hovering, especially in the evening, in search of rodents, or perched high on exposed wires or dead trees. Found in open country, grassland, young oil-palm and the edges of older oil-palm plantations, forest edge next to cultivation. Occurs from India through south-east Asia to New Guinea; resident.

BRAHMINY KITE *Haliastur indus* 45cm

Though a scavenger, this kite is one of the more spectacular coastal birds, often soaring in thermals in large numbers. Adult reddish-chestnut with white head and upper breast. Juvenile dull brown with pale patch under wing at base of primaries. In flight, long but broad angled wings and long square-ended tail distinguish from the larger sea-eagle and smaller, slimmer harriers. Call a thin cat-like mew. Feeds on mudflats and open ground, especially on carrion and small prey. Found typically near coasts, especially mangroves, occasionally inland and on big rivers. Occurs in south-east Asia down to Australia; a common resident, especially on the west coast.

WHITE-BELLIED SEA-EAGLE *Haliaeetus leucogaster* 70cm

This is the biggest raptor in the area (other than the odd very rare vagrant vulture), with an impressive wingspan. Adult has pure white head and underparts, pearl-grey to brownish-grey upperparts, with contrasting blackish primaries and secondaries. Juvenile dark brown, with paler head, throat and base of tail, and dark primaries (pale bases) and secondaries. Confusing intermediate stages of increasing paleness; identification is then aided by size and shape. Loud serial screaming; catches fish from water surface and sometimes feeds on refuse. Found mainly on coasts, wooded rocky shores and mangroves, sometimes inland. Occurs from India to Australia; resident.

23

CRESTED SERPENT-EAGLE *Spilornis cheela* 54cm

This is the most commonly seen forest raptor, a robust, heavy-headed, chocolate-brown eagle. Adult has very short, square black and white crest, white spots on breast; in flight broad, rounded wings and tail with heavy black and white bars bolder than other raptors. Juvenile has pale head and underparts, becoming progressively darker with age. Yellow legs and cere. Call a piercing *a cheee chee!* especially in display flight. Known to take reptiles, small mammals, pigeons. Found in lowland and hill forest, soaring over forest and adjacent open country. Occurs from India and south China to Borneo and Java; resident.

GREATER SPOTTED EAGLE *Aquila clanga* 70cm

This is a speciality of the west coast down to Singapore, seen in small numbers each year. Adult a big all-dark eagle with pale under tail-coverts; broad wings with wide bases and short rounded tail. Yellow legs and cere. Juvenile similar but heavily spotted with rounded white marks that form bars across upper surface of wings, some white on upper tail-coverts. Call a dog-like bark, *chuck chuck*; feeds on small ground mammals, injured birds; often settles on ground as well as in dead trees. Found in open country, swamps, especially along coast. Occurs through most of Europe and Asia to Thailand, Indochina; occasional migrant.

CHANGEABLE HAWK-EAGLE *Nisaetus limnaeetus* 65cm

A most variable bird of prey, some individuals are nearly white, others nearly black. A large raptor with very broad wings, long round-ended tail; no crest. Dark phase birds blackish-brown, with pale flight feathers on underside of wing; grey under base of tail; thus distinguished from the Black Eagle *Ictinaetus malayensis*. Light phase birds dark brown above, nearly white below; several narrow bands on tail and wings (juvenile Crested Serpent-eagle has broader, darker bands). Call a high quavering *yee-ee-ee-ee*. Found in plantations, logged forest. Occurs throughout south-east Asia from India to the Philippines; resident.

BLYTH'S HAWK-EAGLE *Nisaetus alboniger* 52cm

A medium-sized but spectacular crested eagle of hills and mountains. Adult black and white, brownish tone to wings, with strongly barred underparts, long crest and broad white band across dark tail. Juvenile dark brown above with pale scales, pale head and underparts, long crest, several narrow dark bands across buff tail. Call a high quavering scream similar to last species; catches arboreal mammals, birds, reportedly even bats; spends long periods quietly perched. Found in hill and montane forest where frequently seen, seldom in lowlands. Occurs only in the Malay Peninsula, Sumatra, Borneo; resident.

CRESTED PARTRIDGE *Rollulus rouloul* 25cm

The commonest forest partridge, forming devoted pairs with strong family life, congregating into flocks. Male very dark, glossy blue with maroon crest, red legs and skin round eye, and bright red patch on bill; head outline distinctive. Blackish brown wings contrast with metallic purplish-blue of body. Female dull grassy green with grey head, rufous wings, reddish legs and eye skin. Forages for insects and fruits in leaf litter, flocks working gradually through forest; call a thin glissading whistle. Found in lowland and hill forest up to 1,500m. Occurs in the Malay Peninsula, Sumatra, Borneo; resident.

RED JUNGLEFOWL *Gallus gallus* 40–75cm

This is the ancestor of the domestic chicken, therefore arguably the most important and valuable bird in the world. Male unmistakably chicken-like but slim, with grey (not yellow) legs, pale powder-puff at base of tail, and typically big white ear-patch. Female small, dark, with ochre-yellow streaks on neck and rufous tail; dark feet. Domestic fowl lack one or other of this combination of characteristics. Call a four-note crow *ka ka-kaa ka!* which ends abruptly; domestic cockerels often drag out last note. Found in plantations, forest edge, secondary growth. Occurs truly wild from north India to Java; resident.

CRESTLESS FIREBACK *Lophura erythrophthalma* 45–50cm

A modestly plumaged (and therefore inconspicuous), thinly distributed pheasant. Male dark blue-black with plain bright buff tail, and (usually concealed) iridescent orange back; facial skin scarlet. Female dark dull blue-black including tail, greyer head, dull red facial skin. Single birds, pairs or small parties move through undergrowth, gurgling and clucking to stay in contact; males whirr wings as display. Found in logged and unlogged lowland and hill forest up to 800m, especially with stemless palm undergrowth. Occurs in the Malay Peninsula, Sumatra and Borneo; resident.

CRESTED FIREBACK *Lophura ignita* 65cm

Found at few sites, but is almost a guaranteed sighting on Taman Negara riverside trails. Male deep shiny blue-black, arched white centre tail feathers, white flank stripes, and brilliant blue facial skin. Iridescent coppery back usually hidden. Female chestnut-brown, scaled whitish below, becoming white on abdomen; dull blue facial skin. Both sexes have red legs and funny bobble crest. Forages for fruits and insects in leaf litter, in noisy clucking flocks and sometimes singly. Found in lowland forest especially near rivers and streams. Occurs in the Malay Peninsula, Sumatra and Borneo; resident.

MOUNTAIN PEACOCK-PHEASANT *Polyplectron inopinatum*
45–65cm

A subdued-coloured bird best found along walking trails in mossy forest. Dark grey, with wings and tail rufous, appearing chestnut in sunlight; tail long and tapering; wings with small blackish eye-spots, tail and tail coverts with oval to elongate green iridescent patches. Female has shorter, duller tail. Usually silent, solitary or small parties, foraging in leaf litter on ridge tops and slopes; male call a harsh not loud *kek-kek* or *kek-kek-kek*, repeated every few seconds. Found in cool, damp montane forest above 800m, including stunted elfin forest. Endemic to Peninsular Malaysia.

MALAYAN PEACOCK-PHEASANT
Polyplectron malacense 40–50cm

Though silent and hard to find, at odd periods (including heavy fruiting months) becomes very noisy and obvious. Male small but heavily feathered, gingery brown with dark crest that points directly forwards over bill, orange face, round-ended (not tapered) tail; iridescent blue-green eye-spots on wings and tail. Female small, dull grey-brown, no obvious crest, dull orange face, obscure blackish eye-spots. Call of male an angry chicken-like clucking, an explosive single shout, and mellow penetrating two-note rising whistle *hoo hui*. Found in extreme lowland forest over level ground, occasionally in foothills up to 200m. Endemic to the Malay Peninsula.

GREAT ARGUS *Argusianus argus* 75–190cm

Spectacular Malaysian bird, always heard and seldom seen. Both sexes dark brown, markings a fine mosaic of buff, brown and black; chestnut breast, dark tail, and bare blue skin on face and neck. Male gigantic because of elongated wing and tail feathers, complex pattern with eye-spots on wings revealed in display. Small inconspicuous crest. Males call from cleaned display space, very loud two-note hoots, *ka-wow!* Males and females give a similar-toned series of about 30 single hoots. Found in lowland and hill forest to 900m. Occurs in the Malay Peninsula, Sumatra and Borneo; resident.

WHITE-BREASTED WATERHEN *Amaurornis phoenicurus* 33cm

This, the commonest of its family locally, advertises its presence with wild eccentric gurgles and clucks for minutes on end. Adult has white face, breast and abdomen, chestnut flanks and dark grey (nearly black) crown and upperparts; greenish legs, mainly yellow bill, and pale rufous under tail-coverts. Juvenile similar but duller, paler, dirtier-looking. Can be seen crossing roads, flying out of swamps or grassland, often accompanied by black fluffy chicks. Found in agricultural estates, secondary growth, ditches, rice fields, swamps and mining pools. Occurs from India to the Philippines; resident and migrant.

WATERCOCK *Gallicrex cinerea* 42cm

One of the few birds in the region to show striking seasonal changes in plumage. Here typically non-breeding plumage: bulky, buffish brown, with dark brown to blackish streaking above and narrow barring on breast and flanks; dull green bill and legs. Breeding plumage male: blackish with brown scaling on wings and back, rufous beneath tail; red legs and mainly red bill and shield on forehead. Many intermediate plumages seen, seldom in full breeding plumage here. Similar range of eccentric calls to White-breasted Waterhen but deep, booming, brief. Found in marshes, rice fields, behind coastal mangroves. Occurs throughout east Asia; migrant, might breed.

COMMON MOORHEN *Gallinula chloropus* 32cm

This striking and familiar bird is found almost throughout the world. Adult slate-grey to black with broken white stripe on flanks and two white patches separated by black line beneath tail; red forehead shield and bill with yellowish tip. Juvenile dull dark grey-brown, dusky whitish below with white patches beneath tail; dull greenish legs and bill. Call a single musical croak; often seen swimming, occasionally walking, often several to many in open water, swimming with raised tails. Found on mining pools, rice fields, swamps. Occurs throughout the world except poles and Australia; resident mainly on the west coast.

PURPLE SWAMPHEN *Porphyrio porphyrio* 42cm

Though big and brightly coloured, Purple Swamphens are secretive, generally quiet, and always occur in only small numbers. Adult unmistakable bright bluish-purple, green on wings, greyer on head; big red bill and legs, white beneath tail. Juvenile duller with blackish bill. Call a varied harsh cackling and grunting, poorly described; seen usually in ones or twos at edge of water in lotus, reeds, grasses; feeds mainly on plant matter, some invertebrates, often holding food in foot. Found in rice fields, swamps, old mining pools, fish ponds, preferably with lilies and Water Hyacinths. Occurs throughout Africa, southern Asia, Australasia; resident.

MASKED FINFOOT *Heliopais personata* 50cm

One of the most charismatic birds, sought after by all birdwatchers for its oddity. An elongated duck-like bird, pallid grey-brown with banana-yellow to orange bill, topped by inconspicuous forehead bobble when breeding. Dark crown and whitish stripe down sides of neck, throat black in male, white in female. Immature birds have pale crown. Call said to be high-pitched bubbling; typically seen swimming, keeping to dark overhung banks. Found singly or in small groups along forest-lined rivers, pools, and in mangroves especially on migration. Occurs from east India to Indochina and Sumatra; non-breeding visitor recorded in almost every month.

PHEASANT-TAILED JACANA
Hydrophasianus chirurgus 30–55cm

These are the lily-trotters famous for their reversed sexual roles, females mating with several males, each of which cares for nest and young. Elegant, elongated, with very long toes and claws. Non-breeding: tricolour appearance, brown back and short tail, white beneath and on spread wing, black wing-tips, crown, neck-stripe and breast-band; buff hind neck. Breeding: mainly black with white wing-patches, face and neck; long tapered tail. Call a high-pitched repetitive nasal note; usually seen here singly, walking delicately over mud. Found in mangroves, coastal swamps, overgrown mining pools. Occurs from India to the Philippines; non-breeding migrant.

BLACK-WINGED STILT *Himantopus himantopus* 38cm

This is a slim and unmistakable aristocrat amongst waders with proportionally the longest legs of any bird. Etiolated, with red legs, black bill, black back and wings. Underparts and most of head and neck white. In flight, solid dark wings and white head and rump, with trailing feet distinctive. Noisy piping *chip chip chip* in flight; typically in small groups separate from other waders, feeding in deep water by wading and swimming. Found in freshwater swamps and coastal mudflats. Occurs through most of the Old World; a rare visitor.

PACIFIC GOLDEN PLOVER *Pluvialis fulva* 25cm

A common species which is considerably larger than the Little Ringed Plover or Lesser Sand-plover. Non-breeding: speckled brown, buff and gold above, appearing brown at distance, with buffish breast, eyebrow, face and buffish, abrupt forehead. Breeding: upperparts become brighter gold, face and underparts black, with white band from forehead over brow down sides of breast to flanks. Intermediate plumages with some black below. In flight, plain brown above (no white), dirty light buff below. Call a shrill *keruit*. Found commonly on short grassland, rice fields, muddy coasts. Occurs through most of Asia; migrant.

LITTLE RINGED PLOVER *Charadrius dubius* 17cm

A hunched and hesitant runner on mud or sand, this little plover has a rapid direct flight low over the ground. Small, with brown upperparts, white below; black mask extending into patch on forehead with white above and below it, white collar and black breast-band; legs and eye-ring yellow. Juvenile has black replaced by brown, overall duller, obscurely marked. In flight small, no wingbar, white edges to tail, call a long-drawn-out, descending whistle. Found on coasts, mudflats and sand bars, short grassland and riverbanks. Occurs from Africa through Asia to New Guinea; migrant and non-breeding visitor, mainly on the west coast.

LESSER SAND-PLOVER *Charadrius mongolus* 20cm

Lesser Sand-plovers (middle three birds here, with Curlew Sandpipers to left and right) are hard to tell from Greater Sand-plover *Charadrius leschenaultii*. This species small with short bill and dark grey legs, sewing-machine feeding action; Greater has robust bill, green-grey legs, and a more deliberate feeding action. Non-breeding: sandy brown above and on patch at sides of breast; white below and on forehead-cum-eyebrow. In flight, narrow white bar along wing, white-bordered tail, call *pipip*. Adopts partial breeding plumage before leaving, with broad rufous breast-band, white forehead with black above it and through eye. Found on muddy and sandy coasts, seldom inland. Occurs throughout the Old World; migrant and non-breeding visitor.

BLACK-TAILED GODWIT *Limosa limosa* 40cm

Tall, long-legged and straight-billed, this is the commoner of two godwits. Non-breeding: greyish brown above faintly mottled, dusky whitish below and on eyebrow, with black-tipped, pink-based bill, black-tipped tail, white rump and (in flight) striking white wingbar. Breeding: becomes rufous on head, neck and breast, barred on flanks and abdomen. Bar-tailed Godwit *L. lapponica* is more mottled above, with barred tail, and almost no wingbar; in breeding plumage more deeply and extensively rufous. Both species occur on coasts and mudflats; Black-tailed also sometimes inland. Occurs throughout the Old World; migrant and non-breeding visitor.

WHIMBREL *Numenius phaeopus* 44cm

More elegant and commoner than Eurasian Curlew *N. arquata*, the Whimbrel is widespread on mangrove coasts. Dark brown, with downcurved bill and strong blackish and buff stripes on crown; in flight it has a pale off-white rump smaller than in the Eurasian Curlew; wing plain above, pale below with heavy dark barring. Call a one-tone multiple trill; common feeder on mudflats, coming to mangroves to roost. Found on all muddy coasts, swamps behind mangroves. Occurs almost worldwide; non-breeding visitor and migrant, mainly on the west coast.

COMMON REDSHANK *Tringa totanus* 28cm

An abundant and familiar wader worldwide. Non-breeding: grey-brown above and on head and neck; red legs and red bill with black tip; distinguished from Spotted Redshank *T. erythropus* by more uniform upperparts and fainter eyebrow. Breeding: darker, increasingly mottled above, heavily streaked below. In flight, broad white patch from inner primaries across entire secondaries, white rump, narrowly barred tail. Gives three-note or four-note piping call with first note most emphatic. Found on muddy coasts, occasionally swamps inland. Occurs throughout Europe, Africa and Asia to Sulawesi; a very common migrant and non-breeding visitor.

COMMON GREENSHANK *Tringa nebularia* 35cm

The second commonest of the larger sandpipers. Non-breeding: light grey mottled above, on head and sides of neck, whitish below, with long green legs and robust, straight, black bill. Breeding: more heavily mottled. In flight, plain wings, white rump extending up back, barred tail beyond which feet project moderately. Superficially like Marsh Sandpiper *Tringa stagnatilis* but large size, robustness, and loud, ringing, three-note call in practice make separation easy; often in big flocks. Found on coastal mudflats, rice fields, swamps. Occurs throughout the Old World and the Pacific; migrant and non-breeding visitor, mainly on the west coast.

WOOD SANDPIPER *Tringa glareola* 23cm

The squared-off head and chequered wings of this freshwater wader are often sufficient to clinch identification. Upperparts grey-brown, strikingly freckled, breast greyish, head dark with pale eyebrow; yellowish legs. In flight, square white rump contrasting with dark back, underwings pale near body, no wingbar. Typically noisy, *chiff chiff chiff* and more varied notes in flight; feeds around freshwater margins on insects and insect larvae. Found in swamps, rice fields, old mining pools, mangroves; one of the commonest freshwater waders. Occurs throughout the Old World; migrant and non-breeding visitor.

TEREK SANDPIPER *Xenus cinereus* 25cm

This is the only sandpiper in the region which characteristically bobs up and down, apart from the much smaller Common Sandpiper. Long, slightly upturned black bill with yellow base, bright yellowish-orange legs; upperparts dark grey faintly mottled, pale eyebrow, rump only a little paler than back; below, white. Wing with broad white panel on secondaries is distinctive in flight; when standing, white panel is concealed, but dark carpal joint may be seen. Call a quick, high-pitched *yee yee yee*. Found on coastal mudflats, typically in small groups or singly, mixed with other wader species. Occurs throughout the Old World when not breeding; migrant.

COMMON SANDPIPER *Actitis hypoleucos* 20cm

Abundant but typically solitary, this species skims low over water on down-bowed wings. Sandy brown above, and on crown, with light brown patches either side of breast; underparts otherwise white, eyebrow pale. In flight, striking white wingbar and white panels confined to sides of tail. Teetering motion with bobbing tail while feeding along water's edge, call a high-pitched, three-note piping on taking flight. Found in dune slacks, river edge, rice fields, swamps, mangroves. Occurs throughout the Old World; common migrant and non-breeding visitor.

RUDDY TURNSTONE *Arenaria interpres* 23cm

This is an attractive, short-legged wader with complex head and wing markings, usually seen in only small numbers mixed with other species. Non-breeding: mottled dark brown on head, back and wings, the breast darker; lower breast and abdomen white. Orange legs, short black bill and abrupt forehead. Breeding: complex pied head and neck pattern, increased plain rufous on back and wings. In flight a white triangle and narrow white wingbar, white back separated by dark rump from white tail tipped black. Trots around on mud flicking over weeds and pebbles in search of food. Found on mudflats. Occurs worldwide; migrant and non-breeding visitor.

PINTAIL SNIPE *Gallinago stenura* 25cm

Half-a-dozen pairs of pin-shaped tail feathers thrum in the display flight of this, the commonest snipe locally. Usually seen in brief, direct flight showing little or no white trailing edge of secondaries, no white on underside of wing, with feet projecting well beyond tail. On the ground, very long-billed sandy brown waders with head stripes and pale braces. Found in ditches, swamps, rice fields, edges of old mining pools in lowlands. Occurs through east and south Asia to Sulawesi; migrant. Common Snipe *G. gallinago* has white beneath and on trailing edge of wing and erratic flight, Swinhoe's *G. megala* has heavier flight; feet project little.

ASIAN DOWITCHER *Limnodromus semipalmatus* 34cm

This wader, like a miniature godwit, is a mouthwatering find for most birdwatchers, picked out by thorough examination of mixed flocks. Non-breeding: dull grey-brown above, mottled like Bar-tailed Godwit but smaller, very long bulbous-tipped bill held downwards; underparts finely mottled grey. Breeding: back browner, face, breast and abdomen rufous. In flight, vague whitish wingbar and rump with fine darker markings. Usually silent; feeds by pivoting body on stiff legs, neck stiffly forwards, deliberately probing deep with bill. Found on muddy coasts and estuaries. Occurs through east Asia to north Australia; migrant and non-breeding visitor in small numbers.

CURLEW SANDPIPER *Calidris ferruginea* 22cm

Massing sometimes in huge flocks, this is one of the more brightly coloured waders. Non-breeding: scaly pale grey above, whitish below and on eyebrow; dark legs and dark decurved bill. In flight, square white rump and narrow wingbar. Breeding: bright chestnut on face and underparts, back scaly brown. Many individuals can be seen in attractive partial breeding plumage before departure each year. Clean, slim, thin-legged appearance and bill shape characteristic. Found on coastal mudflats, estuaries, ponds and occasionally swamps inland. Occurs through most of the Old World; non-breeding migrant.

LITTLE TERN *Sternula albifrons* 22cm

Its small size, narrow wings and typically sandy beach habitat are good identification guides. Tiny, grey above and white below, with a black cap separated from bill by white forehead. Breeding: feet yellowish, bill yellow with black tip. Non-breeding: feet and bill blackish. Juvenile similar but some obscure scaling above and dark patch on carpal joint. In flight tiny, often hovering, black panel formed by outer primaries. Found on sandy beaches, estuaries, mudflats, coastal ponds. Occurs almost worldwide; here resident and non-breeding migrant, widespread but numbers generally small.

GREAT CRESTED TERN *Thalasseus bergii* 45cm

This is a large but slim tern; back and rump grey, underparts white. Black cap when breeding, blackish mark behind eye to nape when not breeding; short bushy crest. Yellow bill and black legs: in flight, obscure dark mark along trailing edge of primaries beneath wing. Found in coastal waters. Occurs discontinuously from the Indian Ocean to Australia; an occasional non-breeding visitor on the west coast. Care is needed in distinguishing the Great Crested from the Lesser Crested Tern *T. bengalensis*, which is slimmer and lighter, with an orange bill and paler grey wings. The very rare Chinese Crested Tern *T. bernsteini* has black-tipped yellow bill and paler wings.

ROCK PIGEON *Columba livia* 33cm

A familiar bird throughout the world, it has become naturalised in a few places but is generally dependent on human feeding. Wild-type birds are bluish-grey with two black wing-bars, black tip of tail, and glossy green and violet patch on sides of neck. There are many colour phases of domesticated stock, from pure white to chocolate and parti-coloured. Call the familiar *koo kooo kuk*. All Peninsular Malaysian and Singapore birds are derived from released stock, continually supplemented. Found in small flocks, usually close to settlements. Occurs almost worldwide; resident.

RED COLLARED-DOVE *Streptopelia tranquebarica* 23cm

Smaller than the preceding species, the Red Collared-dove has a simpler pattern but brighter colours; nevertheless caution is needed in identifying it at a distance. Small size; grey head with simple black half-collar, vinous pink breast and deep vinous upperparts; conspicuous whitish corners to tail in flight. Female duller, more grey-brown. Call is a repetitive stuttering coo. Found in secondary growth, open country, perhaps preferring sites near the coast. Occurs from India to the Philippines; here it is resident (perhaps introduced) in northern Singapore, previously also Malacca and the Perak coast.

41

SPOTTED DOVE *Streptopelia chinensis* 30cm

Feeding on the ground, this is the most abundant pigeon of open country and cultivation. Brown above, mottled darker brown; vinous pink head and underparts, with white-spotted black half-collar joined round sides and back of neck. Taking off, conspicuous whitish sides to tail and pale grey panel on carpal joint of each wing. Singly or in pairs on ground, feeding on seeds and other bits of vegetation, or perched on wires or low trees; not large flocks. Call a three- or four-note *coo*. Found in oil-palm and rubber, scrub, secondary forest, villages and gardens. Occurs from India through south-east Asia, introduced through to New Zealand and in America; resident.

LITTLE CUCKOO-DOVE *Macropygia ruficeps* 30cm

Heard more often than seen, this is the commonest of all pigeons in montane forest. A slim, brown, long-tailed pigeon with rufous head, faintly or not barred above and below. Male has green iridescence (hard to see) on sides of neck; female has some dark scaling on breast and upper back. Call a rapid *wuck wuck wuck*, resolving at close range into *kuwuck kuwuck kuwuck*, about two notes per second, for minutes on end. Feeding on small fruits at forest edge, nests often found in dense, low growth such as ferns. Found in montane forest, sometimes down to 500m or even in the lowlands. Occurs throughout south-east Asia; resident.

EMERALD DOVE *Chalcophaps indica* 25cm

A dark green, stubby pigeon flying low and direct over the road in rural areas is likely to be this species. Seen well, it is finely coloured, with iridescent green on back and wings, rich vinous pink on face and underparts, and two whitish bars across rump. Male is overall brighter than female with ashy crown and clear white eyebrow; red legs and bill. Call a soft *tik-cooo*, repetitive, first note inaudible at a distance. Found on the ground, singly or in pairs, feeding on fruit and seeds in oil-palm and rubber plantations, in secondary growth and in lowland forest up to 1,200m. Occurs from India to Indonesia; resident but with long-distance movements.

ZEBRA DOVE *Geopelia striata* 20cm

This very small dove can on occasion be delightfully tame, flying off when the observer is only a metre or two away from it. Pale grey-brown all over, palest on head, the plumage barred and scalloped with narrow black lines. Long, tapered tail with white corners seen on take-off. Feeds on the ground, singly or in pairs, on sandy ground; call a high, far-carrying *ka-do-do-do-do* or *kaddle-a-do-do-do-do*, making it a popular competition cage-bird. Found on road margins, in gardens, agricultural estates, grassland and coastal scrub. Occurs from Burma and Sumatra east to Bali; resident, heavily trapped for trade.

43

NICOBAR PIGEON *Caloenas nicobarica* 40cm

Like the camel, this incongruous bird might have been designed by a committee. Dark iridescent green, becoming blackish on head and wings, with a white tail; the green feathers of the forequarters lengthened into pointed golden-green hackles: knob at base of bill; short, ugly reddish legs. Juveniles are duller, with dark tail and no hackles. A wary bird, feeding on the ground and occasionally in trees. Found on small offshore islands off both east and west coasts. Occurs from the Andamans to the Solomon Islands; here it is resident, not yet recorded from the mainland but presumably wanders between islands.

LITTLE GREEN-PIGEON *Treron olax* 20cm

The smallest green-pigeon is also one of the most intensely coloured. Both sexes have a dark tail with broad grey tip. Male has grey head, orange breast-band, extensive maroon wings and back. Female has a grey head, pale throat and green breast. Distinguished from other green pigeons by small size, small dull bill. Call a soft undulating or lilting *coo*, rather brief. Feeds in small flocks on small ripe fruits including figs, at forest edge and in forest. Found from lowlands to 1,100m in forest and secondary growth. Occurs in the Malay Peninsula, Borneo, Sumatra, Java; resident, but travels long distances.

PINK-NECKED GREEN-PIGEON *Treron vernans* 27cm

This chubby pigeon can be enormously common in coastal scrub when fruits are abundant. Distinguished from other green pigeons by grey tail with black band and grey tip. Male has grey head passing through pink to orange lower breast; green back and wings. Female dull green without markedly pale throat, best identified by tail pattern and association with distinctive male. Call a lilting *coo*, varied and prolonged. Found in flocks, especially near coast in mangroves, scrub, secondary forest, forest edge. Occurs from southern Burma to Sulawesi and the Philippines; resident but travels long distances.

THICK-BILLED GREEN-PIGEON *Treron curvirostra* 27cm

Sometimes conspicuous at fruiting trees in the lower zone of montane forest, this pigeon seldom reaches summits and occurs down to the coast. Bright yellow-green bill with red base, red feet, blue-green skin round eye. Both sexes have grey cap and grey tip beneath tail. Male has maroon wings and back (green in female) and chestnut under tail-coverts (streaked in female). Flocking, feeds on figs and other fruits; look for bright thick bill and eye. From mangrove coasts up to 1,100m in lower montane forest, fairly common inland. Occurs from the Himalayas to the Philippines and Java; resident.

JAMBU FRUIT-DOVE *Ptilinopus jambu* 27cm

Brilliantly coloured, the male is saved from garishness by the marvellous pink-flushed ivory breast. Male dark green above, carmine on head, ivory white below with pink centre to breast. Female duller dark green all over but for dull carmine face, pale abdomen, buff beneath tail. Both sexes have orange-yellow bill, red legs, pale ring round eye. Despite brilliance, often quiet and inconspicuous. Males sway from side to side in display, not bowing. Found in primary and secondary forest from lowlands to low hills up to 1,100m. Occurs in the Malay Peninsula, Borneo, Sumatra; resident but travels long distances.

GREEN IMPERIAL-PIGEON *Ducula aenea* 42cm

Imperial-pigeons are big, bulky birds that swallow large fruits whole, void and thus disperse seeds up to the size of a nutmeg. Soft ashy grey with chestnut under tail-coverts and all-dark tail; green upperparts are not striking. Red feet and red base to grey bill. Now reduced by hunting to few sites (Endau Rompin, various islands, mangrove coasts) especially riverine forest, feeding high in canopy in deeply cooing small flocks. Found in lowland forest, mangroves, coastal areas. Occurs from India to south China and New Guinea; resident but local, disrupted by habitat fragmentation.

MOUNTAIN IMPERIAL-PIGEON *Ducula badia* 46cm

The second biggest pigeon here, not uncommon far from mountains. All-grey with pale chin, maroon back, cream (not chestnut) under tail-coverts and dark tail with broad grey band on tip. Identifiable even in flight by size and tail band; call a distinctive very deep *whomp whoomp*, repetitive, the second note deeper and louder. Solitary, or in small parties feeding in canopy or flying high over forest; display flight flapping steeply upwards to stall and glide down. Found in montane forest, often dispersing to lowlands or even the coast daily to feed. Occurs from India to Borneo, Java; resident with long-distance movements.

PIED IMPERIAL-PIGEON *Ducula bicolor* 40cm

This distinctive coastal and island pigeon is a target for hunters but can still occur in huge flocks. Creamy- to ivory-white, often sullied by food, with black flight feathers and tip of tail; bill and feet blue-grey. Juveniles are greyer. Forms big breeding colonies on offshore islands, where fig trees are abundant; noisy clucking *hoo-hoo-hoo*. Also seen singly or in small groups, especially at dawn and dusk over the coast. Found on small islands, in mangrove and other coastal forest. Occurs from south Burma to New Guinea; resident, but with long-distance movements and interchange between islands.

BLUE-CROWNED HANGING-PARROT *Loriculus galgulus* 14cm

This tiny acrobatic hanging-parrot seeks fruit at the tips of twigs, and is said even to sleep inverted. Very small, brilliant green with scarlet breast and rump patches (puffed out in display), orange-yellow back and black bill. The blue crown is inconspicuous. Female (above) green with black bill and red rump; faint indications of orange-yellow back may be visible. Typically in small groups, high up in canopy, moving from tree to tree. Found in lowland forest up to 1,100m. Occurs in the Malay Peninsula, Borneo, Sumatra; resident.

LONG-TAILED PARAKEET *Psittacula longicauda* 42cm

Parrots are represented in the region by few species: this may be related to fruiting characteristics of the forest. This is the most widespread. Sage green, bluer on back, wings and long narrow tail. Male with red bill and face, black chin; female duller with dull brownish bill, dull red face, blackish chin and line through eye, no blue on back. Flies rocket-like over trees, giving single screeches; underside of wings yellow. Feeds in small parties. Found in oil-palm estates, lowland and hill forest, secondary growth. Occurs from the Andamans through the Malay Peninsula, Borneo, Sumatra; resident.

BANDED BAY CUCKOO *Cacomantis sonnerati* 22cm

Most difficult to spot, this bird's call is the best guide to its presence. Small cuckoo rufous brown above, on crown and behind eye; whitish eyebrow and underparts, overall finely barred blackish. Rather long bill, fineness of barring and frequently erect posture help identification. Call a quick, four-note *tea-cher-tea-cher*, the first and third notes the more emphatic. Found in inland forest and coastal dryland forest, secondary growth and broken forest fragments interspersed with scrub and grassland; mainly lowlands, sometimes hills. Occurs from India to the Philippines; resident.

ASIAN KOEL E*udynamys scolopaceus* 42cm

A remarkable case of rapid spread: in ten years koels have colonised the whole peninsula, as nest parasites of crows and mynas. Big and long-tailed, the male is glossy black all over with red eye and pale heavy bill. Female similar in shape but dark brown with bold spots and bars all over. Calls unmistakable, often starting before dawn: ten or more increasingly loud glissading notes, koel! Also a harsher quick bubbling call, similar tone, *kwow-kwow-kwow-kwow*. Found in mangroves, secondary growth, gardens, town parks. Occurs from India to Australia; here formerly migrant, now resident and migrant.

CHESTNUT-BELLIED MALKOHA *Rhopodytes sumatranus* 40cm

One of the more soberly plumaged of the malkohas, this bird can be tracked down by its slow ticking or clucking from dense vegetation. Dark grey all over but for dark dull chestnut abdomen and under tail-coverts, and white tips to tail feathers; greenish bill and red skin round eye. Black-bellied Malkoha *R. diardii* is smaller, without chestnut abdomen: Green-billed is bigger, paler grey, without chestnut abdomen; red eye-skin outlined by white. Forages in middle storey, giving spaced single ticks and soft mewing calls. Found in forest, forest edge, secondary growth and mangroves. Occurs from south Burma to Sumatra and Borneo; resident.

GREEN-BILLED MALKOHA *Rhopodytes tristis* 55cm

This is the longest of the malkohas, with the longest tail, but it is poorly named as several species have green bills. A big, floppy grey bird, palest on face and becoming progressively darker grey to rear, dull iridescent green on wings and on very long tail, which has broad white tips. Black-bellied Malkoha *R. diardii* is much smaller, darker grey, and has a shorter tail with narrow white tips; Chestnut-bellied is distinguished by chestnut abdomen and under tail-coverts. Found in lowland and hill forest, secondary vegetation, dryland growth behind mangroves. Occurs from the Himalayas to Sumatra; resident.

RAFFLES' MALKOHA *Rhinortha chlorophaeus* 33cm

This small bird, the smallest of the malkohas recorded in Peninsula Malaysia, is unusual in the degree of difference between male and female, and in its mournful call. Male bright gingery brown with barred blackish tail, each feather tipped white; blue-grey skin round eye. Female similar but light grey head, tail bright rufous brown, each feather tipped black and white. Call three to six thin slow notes, descending scale. Feeds in middle storey, peering under leaves for big insects. Found in lowland forest, secondary growth. Occurs from south Burma to Sumatra and Borneo; resident.

CHESTNUT-BREASTED MALKOHA
Zanclostomus curvirostris 45cm

This is the most attractive of the malkohas, its rich colours emphasised by the brilliant eye and bill. Iridescent dark green above, dark tail without white feather tips, and underparts entirely dark chestnut; red facial skin and heavy greenish bill. Seen singly or in pairs, creeping through dense vegetation, gradually working upwards within one tree then gliding down to the next; spaced regular ticking or knocking call. Feeds on winged insects, which it frequently catches on the wing. Found in middle storey of lowland forest to 900m, plantations, secondary growth and gardens. Occurs from south Burma to Java; resident.

GREATER COUCAL *Centropus sinensis* 52cm

Lumbering and archaeopteryx-like coucals have a charm of their own, enhanced by their haunting voice. This species is big and heavy-moving. Bright rufous back and wings, rest of plumage black including head, underparts and long, loose tail. Heavy, horn-coloured bill. Juvenile similar but finely barred grey on black areas, and barred blackish on wings and back. Call a deep, far-carrying *hoo* or *boot*, from a few to many notes, delivered slowly, first descending then gradually rising in pitch. Found in scrub, grassland, secondary vegetation, edges of mangrove, riverbanks. Occurs from India to the Philippines; a common resident.

LESSER COUCAL *Centropus bengalensis* 37cm

This is a smaller version of the Greater Coucal, but by no means a miniature. Same colour pattern of black plumage with chestnut back and wings but typically with some pale streaking on plumage. Small bill, smaller and less lumbering appearance. Juvenile rufous brown, paler below, with many pale buff streaks above and below; tail barred. Intermediates are common. Call a harsh, grouse-like double note, *kok-kok*, repeated; and a short series of *hoop* notes delivered quickly and descending in pitch. Found in scrub, grassland, secondary vegetation, swamps. Occurs from India to east Indonesia; resident.

COMMON BARN-OWL *Tyto alba* 33cm

The owl has spread by utilizing the huge supply of rats in oil-palm estates. Whitish owl, sandy buff and grey above, white tinged with buff below, males darker and more spotted below: heart-shaped white face, brown eyes. Bigger than related forest-living Oriental Bay Owl *Phodilus badius*, greyer and less gingery above, less vinous below, and face not shaped up into ear tufts. Often perches erect, looking slim and long-legged. Sometimes disturbed into flight by day. Found in oil-palm and other plantations, towns, roosting and nesting in old buildings and caves. Occurs almost worldwide; here formerly vagrant, now resident.

COLLARED SCOPS-OWL *Otus lettia* 21cm

Scops-owls are confusing because they occur in various colour phases; this is the most likely to be heard and seen. Dull brown or greyish brown, with pale buff half-collar on rear of neck; short ear tufts and striking buff eyebrows above brown eyes. Below, warm buff with short streaks. The call consists of a soft single *hooup* repeated about every 12 seconds; begins soon after dusk. Feeds mainly on insects and small rodents; perches in the middle storey. Found in secondary vegetation, oil-palm and rubber estates, logged and unlogged forest. Occurs through most of east Asia to the Philippines; resident.

BARRED EAGLE-OWL *Bubo sumatranus* 45cm

Eagle-owls are not rare in forest; this is one of the smaller of the eagle-owls worldwide. A heavy owl, with brown eyes and yellowish bill set in a pale face-mask; narrow black barring on otherwise whitish underparts, becoming brown on breast; mottled brown above and on crown. Head often looks flattish owing to elegant sideswept ear tufts. Juveniles very pale, almost white. Call a deep, double hoo hoo, and a harsh quacking *kakakakak*. Found in logged and unlogged forest in lowlands, seldom far up in hilly land, sometimes near rural villages. Occurs from south Burma to Java and Borneo; resident.

BUFFY FISH-OWL *Ketupa ketupu* 45cm

This mad-looking owl with penetrating yellow eyes comes down to the ground and may be dazzled by car headlights on roads. Rufous with some buff spotting and black streaks above, reddish buff with narrow black streaking below; large ear tufts, and eyes brilliant yellow. Both local name (*ketumpok ketampi*) and scientific name are based on the wild, ululating four-note call. Feeds on fish, frogs, other small animals; footprints sometimes seen in sand at edge of water. Found in any forest near rivers, in tall secondary vegetation, agricultural estates. Occurs from west Burma to Indochina and Java; resident.

SPOTTED WOOD-OWL *Strix seloputo* 48cm

The two wood-owls have characteristic gingery faces, but differ in the overall tone, in markings and their habitat. This one has a greyish bill and dark brown eyes set in a pale rufous face-mask. It is whitish below with spaced narrow bars; dark brown above with rounded white spots on crown, back and wings. The call is a single strong hoot, that sometimes has a quavering introduction; also gives deep growls. It is active early in the evening. Found in oil-palm and rubber plantations, scattered trees, mangroves and secondary forest. Occurs from south Burma to Indochina and Java; resident.

BROWN BOOBOOK *Ninox scutulata* 30cm

The lack of a proper face-mask is characteristic of this medium-sized owl. Plain dark brown head and upperparts, with a rather long, broadly barred tail; whitish below with wide, heavy brown streaks; the eyes are yellow, with pale fluffy feathering at the base of bill between the eyes. Easily identified by call, a rapid, repeated *kewick kewick kewick*, often for several minutes on end. Found in unlogged and logged forest, mangroves, secondary vegetation, sometimes rubber or oil-palm estates. Occurs through most of east Asia, from India to Sulawesi; here resident in forest and mangroves, migrant in other habitats.

55

BROWN WOOD-OWL *Strix leptogrammica* 55cm

This is the biggest Malaysian owl, and observers sometimes jump to the conclusion that it is an eagle-owl because of its size. Distinctive dark brown eyes surrounded by dark smudges, set in deep rufous face-mask; underparts pale but finely and closely barred blackish, breast dusky. The upperparts mottled dark brown; no pale spots on crown. Short eyebrows and throat pale, buffish. The call consists of four notes, the first being the strongest, and variations on this theme. Found in lowland and hill forest. Occurs from India to Borneo and Java; resident.

LARGE-TAILED NIGHTJAR *Caprimulgus macrurus* 30cm

The camouflage of this ground-nester makes it hard to spot amongst scattered dead leaves on open ground. Finely barred and mottled greyish brown all over, with pale moustache, white throat patch, and broad ashy grey eyebrows separated by dark centre crown. Male has white patch on centre of primaries and white corners to tail; these are duskier, inconspicuous in female. Call a loud *klok*, like knocking on wood, slowly repeated many times, often in bouts of two or three with pauses. Found in open country, agricultural and mining land, scrub, waste ground. Occurs throughout south Asia to Australia; resident.

GLOSSY SWIFTLET *Collocalia esculenta* 10cm

Representative of a group of small swifts, this species commonly nests on any buildings, making a cup nest of plant material that is glued together with saliva. Very small, all black (glossy above when seen close to) except for dusky whitish abdomen; tail forked very slightly. No other swiftlet has a pale belly. Hawks for small insects in flight, keeping close to vegetation surfaces and trees. Nests in old abandoned buildings, tunnels and light cave entrances. Found over forest, secondary growth, gardens and other habitats up to 1,500m. Occurs through south-east Asia down to Australia; resident.

HOUSE SWIFT *Apus affinis* 15cm

This most common and noisy of urban swifts may form huge, screaming flocks above old city buildings. Black, the flight feathers browner, with white rump and throat; tail with shallow fork, often spread so fork not seen. The combination of white throat and rump is shared only by Fork-tailed Swift *A. pacificus* which is bigger, with longer and more pointed wings, a longer and deeply forked tail, and flies faster. In flight, a shrill screaming or chattering; nests on buildings, under bridges and on cliffs. Found in all open habitats, from city centres to hill tops, coasts and offshore islands. Occurs from south Europe and Africa to the Philippines; resident.

DIARD'S TROGON *Harpactes diardii* 30cm

Although this is one of the bigger trogons locally, its size is often difficult to judge. The male has black head, throat and upper breast, separated from scarlet breast and cinnamon-brown back by poorly visible pink line. Female has brown head, back and breast, dirty pink abdomen; in both sexes the tail is long, black above, largely white below with black speckles and vermiculations. Perches in lower and middle storey of the forest, taking short flights to new perches, turning its head slowly in search of insects; its call is four mournful, descending notes. Found in lowland forest. Occurs in the Malay Peninsula, Sumatra and Borneo; resident.

SCARLET-RUMPED TROGON *Harpactes duvaucelii* 24cm

This is the smallest trogon in the area, with brilliant facial skin in the male. Male has black head, scarlet breast and abdomen, a cinnamon-brown back and a bright, large rounded patch of scarlet on the rump; no pink line bordering black of head. Bright blue skin above eye, deeper blue bill. Female has dirty pink rump and upper tail-coverts, pink abdomen and under tail-coverts; in both sexes the tail is white below, without speckles. Perches in lower and middle storey, turning head slowly to look for insects; call a rapidly descending series of about a dozen notes. Found in lowland forest. Occurs in the Malay Peninsula, Sumatra and Borneo; resident.

COMMON KINGFISHER *Alcedo atthis* 18cm

Though not particularly common here, this kingfisher the smallest of open country kingfishers is familiar throughout Europe and Asia. Blue-green on crown, moustache, wings and back; shining blue stripe down back; rufous on ear coverts, breast and abdomen. Black bill, tiny red feet. Flight direct, low, showing blue back stripe. Call a simple, high piping, repeated two or three times; not useful for identification. Found at water's edge in open-country swamps, old mining pools, aquaculture ponds, mangroves. Occurs throughout Old World except Australia; resident in lowlands and occasionally at higher altitudes.

BLUE-EARED KINGFISHER *Alcedo meninting* 15cm

This small, deep blue kingfisher nests in the earth banks of forest streams, and is the forest equivalent of the open-country Common Kingfisher. Like that species but deeper blue, with blue (not rufous) ear coverts, deeper rufous underparts, deeper but still brilliant iridescent blue line down back to rump best seen in flight. Bill black with some reddish colour at base, tiny feet red. Solitary, best seen at nest sites. Found in lowland forest up to 900m, commoner in the level lowlands, and maintaining a foothold in overgrown plantations near forest. Occurs from India to Java and the Philippines; resident.

STORK-BILLED KINGFISHER *Pelargopsis capensis* 37cm

The vast bill of this kingfisher makes a formidable weapon for dealing with crabs and fishes. Bill red; head brown becoming light rufous on breast, abdomen and hind neck; blue wings and brighter blue back; longish tail. In flight, big red bill and plain blue wings. Call a loud *kow-koo*, quite pleasant, or a harsher variant when these two notes extend into cackle, repeatedly. Found near coasts, mangroves, agricultural land near sea including rice fields and coconut plantations; larger rivers extending into forest. Occurs from India to Sulawesi; resident.

WHITE-THROATED KINGFISHER *Halcyon smyrnensis* 27cm

Noisy calls make this the most obtrusive of kingfishers, large and conspicuous in rural and suburban areas. Red bill and feet; brown head, breast, abdomen and carpal joints of wings, with white chin and upper breast; blue back, wings and tail. In flight blue with black wing-tips and white patch at base of primaries. Call is a loud harsh *kek kek kek*... trailing off. Found in all rural areas, villages, towns and parks, also in mangroves, near coasts, along rivers and on agricultural land, often on roadside wires and fences. Occurs from the Middle East to the Philippines; resident.

BLACK-CAPPED KINGFISHER *Halcyon pileata* 30cm

More richly and subtly coloured than the White-throated, this species often appears mauve when in flight. Red bill and feet; black crown and sides of head, sharply cut off from buffish white collar and throat, becoming fulvous on abdomen; blue back and tail, black carpal joints. In flight purplish-blue with black wing-tips and white patch at base of primaries. Call similar to White-throated Kingfisher. Found mainly in freshwater habitats, forested rivers, reservoirs, swamps, also mangroves. Occurs throughout east Asia from India to Sulawesi; migrant.

COLLARED KINGFISHER *Todirhamphus chloris* 24cm

The newly created open habitats of the lowlands are being colonised by a succession of previously coastal birds, of which this is one of the most recent. Blackish to grey bill and feet; turquoise-blue crown, back, wings and tail, with white underparts and broad white collar bordered by a narrow black line. In flight, wings plain greenish-blue. Call a harsh *kek-kek, kek-kek*, like White-throated Kingfisher but with the notes in couplets, higher pitched. Found in mangroves, coasts, sandy beaches, and some distance inland in gardens and on agricultural estates. Occurs from the Middle East to the western Pacific; resident.

BLUE-THROATED BEE-EATER *Merops viridis* 27cm

Only bird to breed in the area and depart in the non-breeding season. Bright chestnut crown and back; rest of plumage bright light green, blue on throat and tail with elongated tail feathers. Distinguished from Blue-tailed Bee-eater by chestnut head and back, and lack of rufous on throat. Chestnut-headed Bee-eater *M. lescenaultii* light chestnut crown and back, yellow throat, dark breast band, no elongated tail feathers. Found in open habitats April to August, and in forest canopy and forest edge all year in small numbers. Occurs from south China to the Philippines, Sumatra, Java, Borneo; resident and partial migrant.

BLUE-TAILED BEE-EATER *Merops philippinus* 30cm

This species replaces the Blue-throated Bee-eater according to season, and can be considerably commoner. Bright sage-green, paler below, shading into blue on rump and tail; black mask through eye, chin yellow and throat with a rufous patch. Two central tail feathers elongated into points. In flight, wings chestnut below; glides more than Blue-throated. Call a shrill *chiwi* in flight. Found in all open habitats, rural areas, mangroves, plantations, scrub and secondary growth. Occurs from India to New Guinea; migrant, some breeding records from the north.

WHITE-CROWNED HORNBILL *Berenicornis comatus* 100cm

A quite extraordinary, pigeon-like booming, tracked down in the forest, may eventually lead to this hornbill. Dull greyish horn-coloured bill, ragged puffy white head, white underparts, tail and wing-tips; back and wings black. Male has white breast, female black. White parts of plumage often stained and untidy. In flight, black with white trailing edge to wing, white head, breast and tail. Call a repeated three-note *huh hoo, hoo*, the first note the least emphatic. Found in lowland forest, in the middle or even lower storey. Occurs from Indochina to Sumatra and Borneo; resident.

BUSHY-CRESTED HORNBILL *Anorrhinus galeritus* 88cm

Though some other hornbills form occasional huge flocks, this is the most habitually gregarious, groups of ten or more keeping in contact by continual seagull-like yapping. Greyish-black, the tail paler grey with a broad black tip; has bunch-shaped, inconspicuous crest, and bare bluish skin on face and throat. Bill and feet blackish; the least colourful hornbill. Feeds largely on fruit; flocks follow fruiting pattern of trees. Found in lowland and hill forest up to 1,200m altitude, keeping to middle and upper storey. Occurs in Sumatra, Borneo and the Malay Peninsula; resident.

WRINKLED HORNBILL *Aceros corrugatus* 85cm

Though similar to the Wreathed Hornbill, this species is considerably rarer. Black, with base of tail black and the rest white stained yellow. Male has white head and neck with short black crest, white throat pouch, reddish casque on top of bill and base of bill. Female has black head and neck, blue throat pouch and skin round eye. Thinly distributed, never in big flocks; call a bark often of two notes. Found in coastal and lowland forest, occasionally in hill forest. Occurs in the Malay Peninsula, Sumatra, Borneo; resident.

WREATHED HORNBILL *Aceros undulatus* 100cm

Huge flocks of Wreathed Hornbills (immature shown) have been seen, more than 2,000 in one evening flying over north Perak, in fleets of 100 or more. Black, with entire tail white or stained yellow. Male has short chestnut crest, yellow throat pouch with black bar sometimes visible, little or no casque on bill. Female has black head and neck, blue throat pouch with black bar, but a little red skin round eye. Call a bark, one or two notes. Found in lowland and hill forest, common up to 1,300m. Occurs from east India and Burma to Sumatra, Borneo and Java; resident but with long-range movements.

BLACK HORNBILL *Anthracoceros malayanus* 75cm

The Black Hornbill occurs in two colour phases. Males are all black with a whitish horn-coloured bill and white corners to the tail, only some individuals have a broad dull whitish eyebrow. Female all black with blackish bill and some reddish facial skin. Best identified by call, a disgusting retching sound, irregular. In pairs or family parties, in middle storey. Found in lowland forest, not so conspicuous or vocal as other hornbills, but this is the species that survives best in forest fragments. Occurs in the Malay Peninsula, Sumatra, Borneo; resident.

ORIENTAL PIED HORNBILL *Anthracoceros albirostris* 68cm

Much confusion over the correct naming of pied hornbills in south-east Asia has arisen because of variations between the sexes and between individuals. Black, with white abdomen and flanks, white wing-tips seen as entire white trailing edge to wing in flight, and variable amount of white on lateral tail feathers. Whitish patches on face, pale horny bill with big casque marked with black. Call a clattering laugh. Found in forest edge, riverine forest, secondary growth in lowlands, especially near coast and on islands. Occurs from India to Sumatra, Borneo, Java; resident.

65

RHINOCEROS HORNBILL *Buceros rhinoceros* 90–120cm

The most famous of the hornbills is also the one most likely to be recognised by its call, though occurring at low density. Black, with white abdomen and tail crossed by wide black band; bill yellow with red patch at base, the casque orange-red with a yellow tip. In flight Great Hornbill and Helmeted Hornbill *Rhinoplax vigil* are the only others with black band on white tail, but in the Rhinoceros Hornbill wings are entirely black. Call *kronk*, differently pitched in male and female, duetting in flight *kronk krank, kronk krank*. Found in forest from lowlands to 1,300m. Occurs in the Malay Peninsula, Sumatra, Borneo, Java.

GREAT HORNBILL *Buceros bicornis* 115cm

Not the longest but probably the biggest, heaviest hornbill of the region, found only in small numbers. Strikingly pied, with black face, breast and back, black wings having two broad white bands across; white neck and abdomen, white tail with broad black band. All white parts of plumage stained yellowish. Bill and flat-topped casque yellow. Call a harsh *kronk*, lower pitched than Rhinoceros Hornbill and not in duet. Found in forest from lowlands to 1,500m, mainly in northern and hillier regions but also down to coast. Occurs from India to Sumatra; resident.

FIRE-TUFTED BARBET *Psilopogon pyrolophus* 27cm

The most common hill-station barbet. Once its call has been recognised, this barbet seems to follow the birdwatcher around in the montane forest. Sage-green above and on tail, paler below; sinuous creamy-yellow band across throat to eyebrow and over crown, separating maroon hind-crown from black forehead and greyish face and throat. Bill heavy, yellowish-green, crossed by a dark line and with maroon tuft of feathers at base. Call a cicada-like, high-pitched whirr, repeated with increasing speed. Found only in montane forest above 800m. Occurs only in Peninsular Malaysia and Sumatra; resident.

LINEATED BARBET *Megalaima lineata* 29cm

A northern speciality, this species is confined to coastal plains in districts with a rather seasonal climate. Green back, wings and tail; entire head and breast heavily streaked with buff on darker background, giving a mealy appearance; yellow feet and skin round eye, big pale bill. Call a repetitive toot of two resonant notes, *ku-took*, second note higher. In low trees, occasionally on wires. Found in secondary growth, villages, coconut and fruit orchards. Occurs from the Himalayas discontinuously to Java; resident, south to Kedah on the west coast and Pahang on the east coast.

GOLD-WHISKERED BARBET *Megalaima chrysopogon* 30cm

The most ubiquitous barbet, extending from forest to gardens, is also the biggest in the region. Green with large yellow cheek patches, front half of crown yellow, and rear half of crown varying from red to red-flecked blue; dull grey throat. Call a loud repeated *ku-took*, more than one couplet per second, often for minutes on end. Also a repeated trill of similar tone, becoming briefer and slower. Found in forest, forest edge, plantations, scattered woodlands, occasionally well-established old rural gardens with big trees. Occurs in the Malay Peninsula, Sumatra, Borneo; resident.

RED-THROATED BARBET *Megalaima mystacophanos* 23cm

Its common name, referring only to the red throat, does little justice to the sparkling array of colours on the head of this species. The male is green, with orange or yellow front half and red rear half of crown (may appear all red), but bright red throat distinguishes this from all other barbets in region. The female is green with green head, red hind crown and patch between bill and eye, may have bluish tinge to sides of face and throat. Call here an irregular *took, took-took, took,* heard in canopy. Found in tall forest and sometimes forest edge in lowlands. Occurs in the Malay Peninsula, Sumatra, Borneo; resident.

BLACK-BROWED BARBET *Megalaima oorti* 20cm

The call of this barbet, easily identified, is one of the most characteristic sounds of the montane forest. Adult green with a combination of yellow throat and black eyebrow; small areas of red on forehead, hind crown and sides of breast. Call a repeated *tuk tuk trrrk*, two short and one long; feeds on figs and other fruits in canopy. Found in montane forest, 900–1,500m, usually in the canopy. Occurs from south China to Sumatra; resident. Golden-throated Barbet *M. franklini* is also montane, with yellow throat, but has grey (not blue) sides of face, and lacks black brow.

COPPERSMITH BARBET *Megalaima haemacephala* 15cm

Though this barbet is typically solitary, groups of more than ten have occasionally been seen at fruiting trees. Adult dark green with streaked buff and green abdomen, and brilliant yellow throat and upper breast crossed by a broad band of scarlet; red forehead and yellow marks above and below eye: gaudy and small, markings stand out on dark head colour. Call a repetitive *choink choink choink...* of metallic tone, more than one note per second. The only brightly coloured barbet of open areas. Found in scattered trees in open country, secondary growth, gardens. Occurs through Indian subcontinent to Java and the Philippines; resident.

SPECKLED PICULET *Picumnus innominatus* 10cm

One of the smallest woodpeckers, this is another of the characteristic montane forest birds but can be hard to find. Very small with bright olive back, blackish head bearing two white stripes, on brow and moustache; underparts speckled black on white. Male has yellowish patch on forehead, which is dark grey in female. Forages on trunks and branches of small trees, seldom in high canopy but often near forest edge, singly or in pairs, drumming or giving short sharp call, *tsick*. Found in montane forest, about 900–1,500m. Occurs from the Himalayas to Borneo; resident.

SUNDA PYGMY WOODPECKER
Dendrocopos moluccensis 15cm

From being a specialist in mangrove and coastal habitats this bird is successfully expanding its range inland. It is a small woodpecker with brown crown, white barring on dark brown back and wings; broad blackish moustachal stripe, white throat and streaked whitish underparts. The male has a small red streak on side of crown behind eye. Singly or in pairs, feeding low on small trees; call is a vibrating trill, not striking. Found in mangroves, secondary growth near coast, occasionally inland and is spreading, for example, in Singapore. Occurs discontinuously from India to Borneo and Java; resident.

BANDED WOODPECKER *Chrysophlegma miniaceus* 25cm

One of several confusingly similar woodpeckers, this one is commonly found in plantations and sometimes enters rural and suburban gardens. Male has red head with bright yellow-tipped crest; upperparts dull greenish, finely banded; underparts dull maroon strongly banded with buff; wings maroon, some banding on primaries. Female similar but sides of head dull. Barring above and below separate it from Crimson-winged Woodpecker. Found in lowland forest, plantations, mangroves, scattered trees in rural areas. Occurs in the Malay Peninsula, Sumatra, Borneo and Java; resident.

CRIMSON-WINGED WOODPECKER *Picus puniceus* 25cm

Another brightly coloured woodpecker, which superficially looks easy but can be confusing to identify. Dark olive green above and below with plain maroon-red wings; no barring above or on wings, a little barring beneath on flanks, sometimes to abdomen and under tail-coverts. Red crown and crest with yellow tip; male has red moustache stripe. Juvenile more barred, may be confused with Banded Woodpecker but has green head. Call a two-note *chee chee*, emphatic first note. Found in lowland forest, secondary vegetation, plantations. Occurs in the Malay Peninsula, Sumatra, Borneo and Java; resident.

GREATER YELLOWNAPE *Chrysophlegma flavinucha* 33cm

A bright yellow-tipped crest is found on so many woodpeckers in the region that it is a poor guide to identification. Adult dull olive-green above, dull greenish-grey below; male has throat and tip of crest bright yellow; primaries banded black and brown but upperparts and underparts plain. In female, throat light rufous. One of the larger and commoner montane woodpeckers, in middle storey, often near quiet roadsides at appropriate altitudes; call a spaced *chup; chup; chrrr*. Found in montane forest about 900–1,500m. Occurs from the Himalayas to Sumatra; resident.

COMMON FLAMEBACK *Dinopium javanense* 30cm

The two very similar species of flameback require careful separation by comparison of size and head pattern. This one is golden brown on back and wings, with a bright red rump and black flight feathers; scalloped black on white underparts. Head boldly striped black and white, with single black moustache stripe (double in Greater Flameback *Chrysocolaptes lucidus*); male has prominent scarlet crest, female crestless with black crown streaked white (spotted white in Greater Flameback). Found in plantations, coconut, scattered woodland in lowlands. Occurs from India to the Philippines; resident.

MAROON WOODPECKER *Blythipicus rubiginosus* 22cm

This fine deeply coloured woodpecker is largely a bird of the undisturbed forest. It generally looks very dark, plain maroon-brown without any barring, maroon on back and wings; male has scarlet patch round sides and back of neck. Small size and ivory-yellow bill. Often alone; call a repetitive squeak when foraging, or a high-pitched descending trill. Found in understorey of primary forest, extensive forest areas, and bamboo in the lowlands; has rarely been recorded in overgrown plantations. Occurs in the Malay Peninsula, Sumatra and Borneo; resident.

PACIFIC SWALLOW *Hirundo tahitica* 14cm

This swallow is widespread, but never occurs in the huge numbers characteristic of the Barn Swallow *H. rustica*. Distinguished from Barn Swallow by smaller size, sullied greyish underparts with no dark band bordering rufous throat; rufous extends onto forehead. Dark under tail-coverts marked with white, tail forked but outer feathers not very long. Juvenile browner above, and less rufous on forehead and throat. Found in all open-country habitats, over mangroves and lowland forest, to above 2,000m, breeding under bridges and other man-made structures. Occurs from India and Taiwan south through to New Guinea and west Pacific; resident.

BLACK-AND-RED BROADBILL
Cymbirhynchus macrorhynchus 25cm

The advent of roadside wires has provided an additional nesting site for a species which builds nests overhanging rivers. Chunky, with black upperparts, red rump (concealed when perched) and deep red underparts crossed by a black breast band; long white streak down scapulars. Bill brilliant cobalt blue above and yellow below. Juvenile similar, dull reddish-buff underparts. Call a variety of harsh chucks and wheezes. Found in lowland forest, overgrown plantations, tall secondary growth, especially along rivers. Occurs from Burma to Indochina, Borneo; resident.

BLACK-AND-YELLOW BROADBILL
Eurylaimus ochromalus 16cm

Frequently heard but seldom seen, this bird has a little clown face and remarkable bill colour. Small, stubby, black and white; black head with complete white collar bordered below by black; pink breast, yellow rump and yellow flashes on black wings. Brilliant turquoise and yellow bill, yellow eye. Call a long trill starting slowly *dee, dee, de de de...*, without the introductory bang heard in call of Banded Broadbill *E. javanicus*. Found in lowland forest to 700m, logged forest, overgrown plantations. Occurs in the Malay Peninsula, Sumatra and Borneo; resident.

SILVER-BREASTED BROADBILL *Serilophus lunatus* 18cm

Many nests of this spectacular broadbill have now been found, overhanging tracks or roads in the mountains. Silky-textured grey on head, underparts and back grading into rich chestnut on wings and rump; eyebrow and flight feathers black. Wing shows brilliant blue, black and chestnut patches, tipped white, clearest in flight; tail tipped white; bill brilliant blue and yellow. Female has silvery white band across her breast. Found in montane forest up to about 800–1,600m, in middle storey and forest edge. Occurs from the Himalayas to Indochina, Sumatra; resident.

LONG-TAILED BROADBILL *Psarisomus dalhousiae* 28cm

The bright colours of Long-tailed Broadbills make them seem almost parrot-like in appearance. Adult brilliant green with yellow face and throat, black cap showing a blue crown and nape, lime-coloured ear tufts. Tail long, graduated, blue; wings black with large blue patches that appear white from below. Bill bright yellowish-green. The juvenile is duller, with green head. Call five or six screeches all of the same tone. Found in montane forest from about 800–2,000m, in canopy and the middle storey. Occurs from the Himalayas to Sumatra and Borneo; resident.

GREEN BROADBILL *Calyptomena viridis* 18cm

The head feathering of this broadbill is built up over the bill like a Roman plumed helmet, almost concealing the bill. Chunky, green, the male almost glowing green with three sharply defined black marks on folded wing; black marks before and behind eye. Female paler green, without the black marks. Call a descending series of short pops, accelerating, like a ping-pong ball being dropped on a table, but melodious. Found in lowland forest, in the middle and lower storeys, and often near rivulets. Occurs in the Malaysian Peninsula, Sumatra and Borneo.

BANDED PITTA *Pitta guajana* 23cm

The Banded Pitta glows blue and orange on the forest floor. Crown and sides of face black with very broad orange (banded) brows; throat white, back brown, darker in the male. Underparts in male deep blue with orange bars across sides of breast; in female golden-buff with fine black cross-bars. Wings dark with broad white tips on wing-coverts. Juvenile light buffish-brown with buff eyebrows, bluish tail, white wing-coverts. Call consists of a short *trrr* falling in tone, and a falling whistle *pow*. Found in lowland forest. Occurs in the Malay Peninsula, Sumatra, Borneo, Java; resident.

76

BLUE-WINGED PITTA *Pitta moluccensis* 20cm

The most widespread pitta across Asia to Australasia. Black head with striking buff stripes at side of crown, white throat and fulvous breast and flanks; bright scarlet abdomen. Back green, grading into blue rump; wings with blue carpal joint. In flight, big white patch on primaries, wings otherwise dark with blue base. Distinguished from resident Mangrove Pitta *P. megarhyncha* by smaller, paler clearer crown stripes, slower call. Call a two-note whistle with emphasis on second note. Found in lowland forest, tall secondary growth, and other habitats when migrating. Occurs from India to the Philippines; migrant.

JAVAN CUCKOOSHRIKE *Coracina javensis* 30cm

This characteristic bird of the hill stations may be seen sitting out on an exposed perch, especially early in the morning. Large, dark grey, with blackish smudge through the eye; paler grey on abdomen, and pale tip to tail visible only from below. Female faintly barred on breast, blackish smudge reduced; juvenile more strongly barred, still fainter smudge confined to area before eye. Distinguished from Ashy Drongo *Dicrurus leucophaeus* by round-ended (not forked) tail. Call a bell-like *kling*; perches erect. Found in montane forest above about 800m, forest edge and low trees. Occurs from south Thailand to Javaand Bali; resident.

77

PIED TRILLER *Lalage nigra* 17cm

Though strikingly plumaged and not rare, Pied Trillers are seen only from time to time. Male black and white with a rather flat-headed profile, striking white eyebrow and face, white wingbar; black above and on white-sided tail, with a large grey rump patch. Rump patch and wingbar visible at rest separate this bird from Ashy Minivet. Female brownish-grey above, paler grey below, with striking white wingbar; faint pale eyebrow; underparts and rump finely barred. Found in gardens, plantations, secondary growth and mangroves. Occurs from the Malay Peninsula to Sulawesi and the Philippines; resident.

ASHY MINIVET *Pericrocotus divaricatus* 20cm

The overall impression is of a demure and crisply clean bird, the white parts of the plumage quite unsullied. Male with bright white face including forehead, dark crown and dark line from bill through the eye; underparts white to very pale grey, back and wings plain grey but revealing a white bar across wing in flight. Tail dark, the outer feathers mainly white. The female has paler grey on crown, less white on forehead. Found in forest, forest edge, riverside and secondary vegetation. Occurs from China to the Philippines, Borneo and Sumatra; migrant (October to April).

GREY-CHINNED MINIVET *Pericrocotus solaris* 18cm

Varying light conditions can make this bird difficult to separate from the Scarlet Minivet *P. flammeus*. Male has grey throat and face that need to be looked for carefully; underparts more orange (but juvenile male Scarlet also has somewhat orange underparts); no separate patch of colour on secondaries. Female grey and yellow as other minivets but throat whitish (not yellow), and no trace of yellowish on entirely grey forehead. Behaviour similar to other minivets, in small parties or more often pairs. Found in hill and montane forest, about 800m upwards. Occurs from the Himalayas to Indochina and Borneo; resident.

STRAW-HEADED BULBUL *Pycnonotus zeylanicus* 28cm

The biggest of the bulbuls has the finest song, delivered by duetting pairs in the early morning. Identified by size and by its rich rufous-buff crown, which appears streaked or furrowed; the blackish moustache and mark through eye, pale throat and underparts. The breast and back have obscure narrow light streaks. Typical of, but not confined to, riverbanks, where the lilting song is richer, more melodious than any other bird. Found in forest and forest edge, rivers, sometimes surviving in small forest remnants, plantations in lowlands. Occurs in the Malay Peninsula, Sumatra, Borneo and Java; resident.

BLACK-AND-WHITE BULBUL *Pycnonotus melanoleucos* 18cm

Of all the bulbuls, this is perhaps the least bulbul-like in its appearance. It is seldom seen by the birdwatcher, and is thought to lead a semi-nomadic existence, feeding over very large forest areas. Adult black with a rounded white patch on the wing; in flight, white shows beneath wing. The juvenile dark olive-brown, rump slightly paler rufous; intermediates occur. Seen singly, seldom in mixed foraging flocks, feeding on small fruits high up in canopy. Found in lowland and hill forest, preferring heath and swamp forest. Occurs in the Malaysian Peninsula, Sumatra and Borneo; resident.

BLACK-HEADED BULBUL *Pycnonotus atriceps* 17cm

This bulbul has iridescent black feathering on the head which lies very close and gives the bird a curiously small-headed appearance. It is rather small, with black head and throat; greenish-olive above and yellow below; olive tail with yellow tip and black subterminal bar. Tail pattern seen well in flight. Close up, iridescence on head and bright blue eye are very attractive. Typically seen in small parties, individuals calling with loud single notes, *tui*. Found in forest and on forest edge, secondary growth, low scrub along rivers and roadsides up to 1,000m. Occurs from east India to Indochina, Borneo, Java; resident.

BLACK-CRESTED BULBUL *Pycnonotus flaviventris* 18cm

There is considerable variation in this species over its range, but the unusual pointed crest is enough for reliable separation from all except Red-whiskered Bulbul. Tall black crest standing straight up on black head, with contrasting pale eye; olive back and tail, yellowish underparts. Distinguished from Red-whiskered Bulbul by lack of face pattern, yellow not white below; from Black-headed Bulbul by crest and plain tail. Call a hesitant repetitive song of five or six notes, the last two repeated. Found in lowland and hill forest, forest edge and tall secondary growth from lowlands to 1,200m. Occurs from India to Indochina and Peninsular Malasia; resident.

SCALY-BREASTED BULBUL *Pycnonotus squamatus* 14cm

This easily identified bulbul tends to be found in hilly country, keeping to the tree-tops and seldom mixing with other species. Small, with black head and contrasting white throat, and black and white scaled breast; distinctive yellow under tail-coverts; brown above, with white tips to outer tail feathers seen in flight. Often solitary, feeding on small fruits in canopy, call a series of high single notes not very distinctive. Found in forest in lowlands and hills. Occurs in the Malay Peninsula, Sumatra, Borneo and Java; resident.

RED-WHISKERED BULBUL *Pycnonotus jocosus* 20cm

The only other bulbul with an upright pointed crest is the Black-crested Bulbul, which is distinctively yellow and occupies different habitats. Red-whiskered Bulbul has black crest, red flash behind eye, and distinctive black and white face pattern with white cheeks. Brown above, white below with black mark at sides of breast and red under tail-coverts. Juvenile similar, but red parts faint or absent. Call a short, varied three- to four-note song. Found in a few rural areas, city outskirts, secondary growth mainly in the north, supplemented by escaped cage-bird populations in Penang, Kuala Lumpur, Singapore. Occurs from India to China and the Malay Peninsula; resident.

PUFF-BACKED BULBUL *Pycnonotus eutilotus* 22cm

This bulbul is under-recorded because its nondescript appearance makes it hard to distinguish from three or four other brown bulbuls. Large, plain brown above and whitish below with greyish tinge to breast and buff tinge to abdomen. Tips of outer tail feathers typically but not always whitish, hard to see; eye reddish-brown. There is a faint crest, giving a slightly peaked profile to hind crown. Call four notes, the first emphatic and the remainder rapid, running together. Found in forest and forest edge, keeping to middle and lower storey. Occurs in the Malay Peninsula, Sumatra and Borneo; resident.

STRIPE-THROATED BULBUL *Pycnonotus finlaysoni* 18cm

This attractive bulbul is found in both lowlands and hills in the north, but southwards progressively confined to hilly country. Upperparts, wings and tail dull olive-brown; forehead, face and throat with bright yellow streaks that fade into narrow creamy lines on pale grey breast; under tail-coverts yellow. Call a rising then descending series of about eight melodious notes; typically feeding in lower storey and fringing vegetation. Found in forest edge, secondary growth and scrub, from about 1,000m downwards, occasionally in extreme lowlands. Occurs from Burma to Indochina and the Malay Peninsula; resident.

YELLOW-VENTED BULBUL *Pycnonotus goiavier* 20cm

It is impossible to avoid seeing this, the commonest of all the bulbuls. Brown above, the head chalky white with a dark crown and dark stripe through the eye; underparts are white, tinged brownish-grey on breast which looks faintly mottled; under tail-coverts pale yellow. Crest often raised into a short peak. Call is a brief bubbling series of notes, the two birds of a pair calling and raising spread wings over their backs in unison. Found in gardens, plantations and other cultivated land, towns, open country of all sorts up to hill stations, and in mangroves. Occurs from south Burma to Philippines and Java; resident.

OLIVE-WINGED BULBUL *Pycnonotus plumosus* 20cm

This is one of about five rather similar-looking brown bulbuls, in which eye colour and slight differences in the tone of the breast, abdomen and wings are important for identification. Eye dull red in adults, brown in juveniles. Plumage dark brown above, greyish-buff below with deep buff under tail-coverts. Ear coverts have faint whitish streaks, and feathers of folded wing have narrow olive fringes that together form an inconspicuous olive panel. No crest, and larger than most other brown bulbuls. Found in secondary growth and scrub, logging tracks, keeping to lower storey. Occurs in the Malay Peninsula, Sumatra, Borneo and Java; resident.

CREAM-VENTED BULBUL *Pycnonotus simplex* 17cm

In Peninsular Malaysia and Singapore, adults are the only brown bulbuls with white eyes, but juveniles, and adults of the same species in Borneo, have brown eyes. Plumage dark brown with pale throat, abdomen and under tail-coverts contrasting more with brown-washed breast than in other species. No distinguishing wing panel. Calls and behaviour not distinctive, a typical brown bulbul feeding on small fruits of quick-growing forest-edge plants. Found in lowland forest up to about 900m, secondary growth and forest edge, low growth along logging tracks. Occurs in the Malay Peninsula, Sumatra, Borneo and Java; resident.

RED-EYED BULBUL *Pycnonotus brunneus* 18cm

This bulbul is distinguished by red eyes that lack a striking yellow rim. It is intermediate in size between Olive-winged and other brown bulbuls; dark brown above, brownish throat and breast with little contrast, slightly paler abdomen and under tail-coverts. No distinguishing wing panel, but wings and tail may look faintly washed with olive. Calls and behaviour not distinctive; a high-pitched trill ascending at the end. This is one of the three commonest brown bulbuls. Found in lowland forest up to 900m, secondary growth along forest edge. Occurs in the Malay Peninsula, Sumatra and Borneo; resident.

SPECTACLED BULBUL *Pycnonotus erythrophthalmus* 17cm

A little less common than the other brown bulbuls, this species is nevertheless a likely encounter on birdwatching trips to lowland forest. Rather small, the eyes red with a bright yellow ring of bare skin; plumage brown, the breast washed with grey, abdomen and under tail-coverts washed with pale buff. No distinguishing wing panel. Sometimes appears slightly bigger-headed than other species, rounded crown. Calls and behaviour not distinctive; feeds on small fruits in low trees and up into forest canopy. Found in lowland forest, logged forest along overgrown tracks. Occurs in the Malay Peninsula, Sumatra and Borneo; resident.

OCHRACEOUS BULBUL *Alophoixus ochraceus* 22cm

Primarily of the mountains, but also inhabits hill slopes and overlaps with two similar species. Crown feathers often erected into a short crest; whitish throat. Grey face, grey-brown back and buffish-grey breast grading into yellowish abdomen (poorly visible in this photo), fulvous under tail-coverts. Distinguished from Grey-cheeked Bulbul *C. bres* by longer crest, almost no yellow on abdomen; from Ashy Bulbul by no dark face smudge, browner breast. Call a harsh little song, singly or often with mixed flocks that move through middle and lower storey. Found in hill forest and mountains. Occurs from the Malay Peninsula through Sumatra, Borneo, Java, Palawan; resident.

HAIRY-BACKED BULBUL *Tricholestes criniger* 16cm

Although its common name misleadingly points to a feature which is virtually impossible to see in wild birds, this bulbul is easily identified. Small; warm brown above, dirty yellowish below, becoming yellower on abdomen; face with broad yellowish patch of feathers round eye and pale throat giving unique face pattern. A few very fine dark hair-like feathers lie over the back. Common in middle and lower storey, usually solitary, occasionally in mixed feeding flocks. Found in lowland and hill forest, forest edge up to about 900m. Occurs in the Malay Peninsula, Sumatra and Borneo; resident.

MOUNTAIN BULBUL *Ixos mcclellandii* 24cm

This is a robustly built bulbul of the mountains, where it must be distinguished carefully from the Streaked Bulbul: both have streaks. This species has a slight disorderly crest; definite white streaks on crown, ear-coverts, throat and upper breast; upperparts olive-brown grading into olive-green wings and tail; breast tinged rufous, abdomen off-white, under tail-coverts yellowish. Call three or more mournful repeated notes, either level or ascending in pitch; feeding in middle storey. Found in montane forest and forest edge, about 900m and upwards, at hill stations. Occurs from the Himalayas to Indochina and the Malay Peninsula; resident.

STREAKED BULBUL *Ixos malaccensis* 22cm

Reasonably common along roads in forested hilly country, this bulbul is submontane rather than truly montane. Distinguished from Mountain Bulbul by rounded crown without crest; faint, narrow whitish streaks on grey throat and upper breast; white abdomen and white under tail-coverts; upperparts dark olive-brown not grading into brighter olive on wings or tail. Call a series of short trills; keeps more to tall forest rather than secondary growth or even stunted elfin forest. Found in hill and lower montane forest about 500–1,000m in altitude. Occurs in the Malay Peninsula, Sumatra and Borneo; resident.

ASHY BULBUL *Hemixos flavala* 20cm

Subject to a good deal of regional variation, Ashy Bulbuls in this area are less distinctively marked and need greater care in identification than elsewhere. Striking white throat often puffed out like that of Ochraceous Bulbul; dark crown and blackish smudge through and below eye; upperparts ashy grey including wings and tail, underparts nearly white. Call a varied and melodious song, compared with harsher brief song of Ochraceous Bulbul; moving in small flocks in middle and upper storey. Found in hill and lower montane forest about 400–1,000m, degraded forest and forest edge. Occurs from the Himalayas to Sumatra and Borneo; resident.

COMMON IORA *Aegithina tiphia* 15cm

The best views of this bird are when it is seen singing in bright sunshine, from the crown of a roadside tree. Adult yellow below, varying from olive-green to nearly black above; wings black crossed by two white bars; tail dark and unmarked. Female slightly duller greenish-yellow, similar pattern. Rump may seem white, especially in flight, owing to long white flank plumes. Call a series of quick deep notes forming a slow trill; also a long descending whistle and a double descending whistle. Found in gardens, secondary growth, plantations, riverside trees, mangroves and forest edge. Occurs from India to Borneo and Java; resident.

GREATER GREEN LEAFBIRD *Chloropsis sonnerati* 20cm

This species has the best song of the leafbirds in the region, enriched by its ability to mimic other birds. Slightly larger than other leafbirds, with a robust bill. Male has elongated black throat patch without a yellow border. Female with yellow throat and ring round eye. In flight and occasionally when perched, blue feathering on carpal joint is visible, but is less extensive than in Blue-winged. Call a varied song reminiscent of Magpie-robin. Found in middle storey of forest, forest edge, tall secondary growth up to 1,200m. Occurs in the Malay Peninsula, Sumatra, Borneo and Java; resident.

LESSER GREEN LEAFBIRD *Chloropsis cyanopogon* 17cm

The bright clear green of leafbirds makes them easy to identify as a group, but the differences between species are confusing and demand care. Lesser Green is rather small without blue on wing. Male has round black throat patch with narrow yellowish border. Female has green throat and no blue in wing. Call rather varied, a four-note whistle with third note the highest; feeding in middle storey and low trees, often in pairs or mixed species groups feeding on small fruits. Found in forest edge in lowlands to 900m altitude, secondary growth. Occurs in the Malay Peninsula, Sumatra and Borneo; resident.

BLUE-WINGED LEAFBIRD *Chloropsis cochinchinensis* 17cm

Three leafbirds – this, the Greater Green and the Lesser Green – all occur together in the lowlands. Both sexes of Blue-winged have turquoise-blue extending from carpal joint to primaries of wing, and less obviously on side of tail. Male has elongated black throat patch with distinct yellow border and overall yellowish tone to head. Female has green throat, like Lesser Green. Feeds on small fruits in the crowns of lower trees. Found in lowland and hill forest and forest edge up to 1,200m. Occurs from India to Borneo and Java; resident.

ORANGE-BELLIED LEAFBIRD *Chloropsis hardwickii* 18cm

This montane species is the most brightly coloured of the leafbirds, but some lowland species also extend upwards into the montane zone. Male slightly duller green than other leafbirds, but extensive black face and throat extending onto upper breast; lower breast and abdomen rich orange; bright blue carpal joints and dark flight feathers. Female all green except for orange tone to abdomen and under tail-coverts. Call a varied, musical song. Found in montane forest from about 1,000m upwards, feeding from the forest canopy down into low roadside trees. Occurs from the Himalayas to south China and the Malay Peninsula; resident.

ASIAN FAIRY-BLUEBIRD *Irena puella* 25cm

This restless and exotically plumaged bird is one of the grand sights of the forest, and common. Male with shining enamelled blue upperparts including rump; black face, underparts, wings and tail; eye deep red. Female plain dark turquoise-blue with dark flight feathers, red eye. Calls include piping *pee-dit*, single notes, and frequent liquid song; seen singly, in pairs or in small parties, moving from tree to tree. Found in lowland and hill forest up to 1,400m, in upper and middle storey especially at fruiting figs. Occurs from India to Borneo, Palawan, Java; resident.

TIGER SHRIKE *Lanius tigrinus* 18cm

This stocky, heavy-billed shrike is found in denser vegetation than the other species, especially in forest edge. Adult has grey crown and nape, rufous-brown back, wings and tail with narrow wavy black bars; black mask through eye, and creamy white underparts. Juvenile like Brown Shrike but face mask dusky, barred, and extensive barring on upper- and underparts; thick bill. Seldom calls; does not characteristically perch in exposed positions. Found in forest, forest edge, secondary vegetation, overgrown plantations, mainly in lowlands. Occurs through east Asia to the Philippines and Sulawesi; migrant.

BROWN SHRIKE *Lanius cristatus* 19cm

The commonest shrike in open country is present from about mid-September to April. Caramel or rufous brown with black mask through eye, emphasised by pale forehead and eyebrow; more rufous on crown and tail, light buff to whitish below. Juvenile has faint and narrow blackish barring on flanks and back. Call a harsh chatter, especially in the early morning, mainly for a few weeks after arrival when setting up territories; perches erect on exposed twigs, fences, wires, catching insects on ground. Found in open country, cultivation, gardens, secondary growth. Occurs throughout east Asia to the Philippines and New Guinea; resident.

LONG-TAILED SHRIKE *Lanius schach* 25cm

The only shrike present all year round is an open-country bird, rather richly coloured, with a long tail. Black crown, nape and sides of face, plain bright rufous back, black wings with small rounded white patch at base of primaries; very long dark tail. Cheeks, throat, centre of breast and abdomen all white, with rufous flanks, under tail-coverts and rump. Juvenile similar but duller with some barring on flanks. Found in rice fields, any open country including parks and canal margins, scrub, grassland, mainly in the lowlands. Occurs from India to Taiwan, the Philippines and New Guinea; resident.

EASTERN YELLOW WAGTAIL *Motacilla tschutschensis* 17cm

The very pale colouring of the juvenile and non-breeding adult Eastern Yellow Wagtails seen here is a surprise to those familiar with the species in its breeding range. Adult olive above, yellow to mixed yellow and white below, with white outer tail feathers; head darker olive with pale eyebrow. Juvenile similar but light brownish-olive above, white to buff below, usually with a few dark specks on breast; eyebrow nearly white. During migration will flock with up to a hundred other birds. Found on short grassland, coastal scrub and bare ground, and marshy areas near and in mangroves. Breeds in Russia, China and Mongolia; migrant.

PADDYFIELD PIPIT *Anthus rufulus* 17cm

Only a few species of pipit occur in this region, but Paddyfield Pipit is widespread and abundant. Warm brown streaked with dark brown above, buff below with more or less faint streaks on upper breast, grading into whitish abdomen, and prominent eyebrow; white outer tail feathers visible in flight. Pink legs; trots on short grassland, standing erect when halted; sings in flight. Call consists of a single *chup* on take-off and a thin *chick-a-chick* when descending. Found on roadside verges, short grass, open agricultural land. Occurs through most of east Asia from India to the Philippines; resident and migrant.

ORIENTAL MAGPIE-ROBIN *Copsychus saularis* 20cm

One of the most familiar garden birds, the Oriental Magpie Robin takes up the role occupied by garden blackbirds, robins or thrushes elsewhere in the world. Male distinctively glossy black and white; head, breast, back, wings and cocked tail black; white wing stripe, outer tail feathers and abdomen. Female similar but dark grey and white, juvenile more mottled than female. Hops on ground with raised tail, and in trees and bushes; a fine varied song. Found in all open country, cultivation, gardens, secondary forest, extending into disturbed forest and up rivers. Occurs from Pakistan to the Philippines, Borneo and Java; resident.

WHITE-RUMPED SHAMA *Copsychus malabaricus* 20–28cm

This widespread forest bird has a rich and varied song, more melodious than that of the closely related magpie-robin. Head, breast and wings entirely glossy black in male; bold white rump and long, graduated black tail with white edges; lower breast and abdomen rufous-orange. Female similar but dark grey and rufous rather than black and orange; shorter tail. Song commonly heard, and single angry *chack* notes given from understorey. Often trapped and caged for its song. Found in lowland and hill forest to 1,200m, overgrown plantations, tall secondary growth. Occurs from India to Java and Borneo; resident.

RUFOUS-TAILED SHAMA *Copsychus pyrropyga* 22cm

Though less often seen than the White-rumped Shama, this species is not rare in suitable habitat. Male dark grey on head, breast and wings, with a white fleck before eye; bold orange abdomen, rump and most of tail; last third of tail dark grey. Female similar but grey-brown without a white fleck before eye, breast rufous-buff with faint dark band, abdomen buff. Distinguished from White-rumped Shama by shorter, rufous and grey tail, lack of white rump. Song much less varied, slow glissading whistles. Found in lowland and hill forest up to 900m. Occurs in the Malay Peninsula, Sumatra and Borneo; resident.

CHESTNUT-NAPED FORKTAIL *Enicurus ruficapillus* 20cm

The repeated long, sweet whistle of this bird is one of the characteristic sounds along lowland forest streams. Largely black and white, with white rump, deeply forked black tail with white edges and tips, and chestnut extending from crown to hind neck (male) or along entire back (female); wings black with white cross-bar, breast white with fine black scales. Distinguished from other forktails by chestnut head colour and small size. Flies repeatedly away from approaching watcher, calling. Found on ground along streams within lowland forest up to 900m. Occurs in the Malay Peninsula, Sumatra and Borneo; resident.

EASTERN STONECHAT *Saxicola maurus* 14cm

Breeding mainly in Russia and China, Eastern Stonechats are scarce in this region but can stand out on low bare twigs. Male has black head, dark brown upperparts mottled with black, pale rump, and white patches on sides of neck and on wings; ochre-yellow below. Female similar but head paler, contrasting less with body; underparts buff. Perches erect on tops of low bushes, dives down; gives clicking scolding calls. Found in open waste land, scrub, disturbed vegetation along riversides in coastal plains. Breeds from Japan to Pakistan, winters south to the Philippines; occasional migrant.

CHESTNUT-CAPPED THRUSH *Zoothera interpres* 15cm

This small, strongly marked but inconspicuous forest thrush has turned up in the most surprising places, even on small islands. Mainly dark slaty grey including back, tail, wings and breast; speckled on face and white abdomen, with two white wingbars or patches; entire crown dark chestnut. Juvenile has a crown streaked rufous and white, rufous breast with black spots, and rufous wingbars. It is secretive and seemingly rare but actually widespread. Found in lowland and hill forest mainly below 900m. Occurs from the Malay Peninsula to Borneo, Java and the Philippines; resident.

HORSFIELD'S BABBLER *Malacocincla sepiarium* 14cm

This is one of the more approachable forest babblers that keeps to the understorey. It is dark brown above, head and particularly cheeks grey, and its throat is white. Underparts pale, breast faintly streaked with grey, becoming fulvous on flanks and abdomen; with dull grey, sturdy legs. Tail slightly longer and bill heavier than the Short-tailed Babbler *T. malaccense*, which has a moustache and a plain breast. Call of three notes, *chap chip chooee*; often singly, in thick undergrowth. Found in lowland forest to about 900m altitude. Occurs in the Malay Peninsula, Sumatra, Borneo and Java; resident.

SOOTY-CAPPED BABBLER *Malacopteron affine* 16cm

One of the least distinctive grey tree-babblers is probably best identified by its call. Off-white below with grey tinge on breast; pale grey face, crown slightly darker grey to blackish, upperparts greyish-brown. Distinguished from similar babblers by lack of rufous on crown, and lack of dark moustache. Call a series of whistles, spaced and deliberate, going up and down the scale, rather human-like; often in mixed flocks with other birds including babblers, or in pairs in lower storey. Found in forest of lowlands and foothills. Occurs in the Malay Peninsula, Sumatra and Borneo; resident.

RUFOUS-CROWNED BABBLER *Malacopteron magnum* 18cm

A species pair is formed by the Scaly-crowned Babbler *M. cinereum* and the Rufus-crowned Babbler, the latter seldom being distinguishable on larger size alone. Slightly bigger, dark brown above; underparts pale, washed grey-brown on breast and flanks, with some obscure streaks on breast; hind crown dark, sharply cut off from the rufous chestnut forehead and fore crown. No black scales on fore crown. Call a series of ascending, then descending, then ascending thin whistles. Found in lowland forest and secondary growth to about 500m, in middle and lower storey. Occurs in the Malay Peninsula, Sumatra and Borneo; resident.

CHESTNUT-BACKED SCIMITAR-BABBLER
Pomatorhinus montanus 20cm

The male and female of this attractive babbler have a close pair-bond and indulge in mutual preening. Head black with a clear white eyebrow, white throat and breast; behind this, from back down the sides of breast, flanks and abdomen, is bright chestnut; chestnut rump contrasting somewhat with dark wings and tail. Bill long, yellow to horn-coloured, curved downwards. Call two or three mellow hoots, also a series of rattling or rasping notes. Found in lowland and hill forest to about 1,200m, in middle storey. Occurs in the Malaysian Peninsula, Sumatra, Borneo and Java; resident.

STRIPED WREN-BABBLER *Kenopia striata* 14cm

This babbler, though not often seen, appears to be common particularly in forest in the extreme level lowlands. White face with a buff patch between bill and eye, white underparts becoming progressively more buff towards under tail-coverts; crown black with white scales or streaks, becoming brown with white streaks on the back; brown tail. Call a three-note whistle *Be my guest*; on and close to ground in low-lying areas, swampy undergrowth. Found in extreme lowland forest, seldom in hills. Occurs in the Malay Peninsula, Sumatra and Borneo; resident.

STREAKED WREN-BABBLER *Napothera brevicaudata* 14cm

This noisy bird sometimes occurs in small flocks, but can still be hard to see in the dark understorey amongst the boulders it favours. Olive-brown above, strongly dappled with pale centres and dark rims to feathers; throat with obscure black and whitish streaks grading into warm fulvous or rufous underparts. Loud shrill whistles are given; a two-note rising *tew-ee*, also three notes and a single descending whistle. Found in forest from lowlands to about 1,500m, limestone hills where forest remains, amongst gullies and rocks. Occurs from east India to China and the Malay Peninsula; resident.

GOLDEN BABBLER *Stachyridopsis chrysaea* 12cm

Scarcely similar to the other *Stachyris* babblers, this bird looks superficially more like a warbler. Overall yellow appearance; black crown with yellow streaks, black triangular bill with black feathers from bill to eye; underparts entirely dirty yellow; wings and short tail olive-brown. Song a series of low whistles, the first emphatic and the rest running together; lurking in ferns and other understorey vegetation. Found in montane forest and forest edge, scrub along roadsides above 900m altitude. Occurs from the Himalayas to south China and the Malay Peninsula; resident.

GREY-THROATED BABBLER *Stachyris nigriceps* 13cm

One of the few birds that is confined to middle altitudes rather than being overtly montane, this babbler is extremely common at the appropriate heights. Plain fulvous brown, rufous-buff below; head pattern rather blurred including black and white streaked crown, white moustache bordered above by black and below by blackish throat; sides of face fulvous. Voice not distinctive but helpful in following flocks, a chattering and buzzing; moves in small parties through dense undergrowth, especially at sides of tracks and roads. Found in hill and montane forest about 500–1,200m. Occurs from the Himalayas to Sumatra and Borneo; resident.

CHESTNUT-WINGED BABBLER *Stachyris erythroptera* 13cm

Typical features of the *Stachyris* babblers include rufous plumage, black feathering somewhere on the head, and bright blue skin about the eye or throat. This species has plain dark grey head and breast, paling onto abdomen; bare blue skin in front of and round eye; chestnut wings and tail. Call a quick series of poop notes, descending in pitch towards end. Found commonly in lowland forest, forest edge and logging tracks to about 900m, in understorey. Occurs in the Malay Peninsula, Sumatra and Borneo; resident. Similar Grey-headed Babbler *S. poliocephala* has less extensively streaked grey head, rufous breast.

PIN-STRIPED TIT-BABBLER *Macronous gularis* 12cm

One of the babblers most likely to be seen in disturbed habitats rather than forest is this light yellowish bird. Small with chestnut crown and yellowish eyebrow, pale yellow underparts more or less thinly streaked with black; back olive-brown, more rufous on the wings and tail. Two familiar calls: a repeated chonk chonk chonk often in clusters of three or four notes, and a harsh *sheet-sheee* with emphasis on the second note. Seen moving in small trees, feeding on insects as it goes. Found in forest edge, secondary growth, casuarina and bamboo in lowlands. Occurs from north-east India to Indochina, Peninsular Malaysia and Sumatra; resident.

SPECTACLED LAUGHINGTHRUSH *Rhinocichla mitrata* 22cm

This laughingthrush is the most likely to be seen, noisy and sociable but elegant in appearance. Ashy grey with chestnut crown, bright white eye-ring and streaks on forehead, and white wing panel formed by pale edges to primaries. Underparts brownish-grey, under tail-coverts chestnut; bill bright orange. Call a repeated two- or three-note whistle *too-tuioo*; moving in small parties, often with other species, in middle and lower storey, especially forest-edge vegetation. Tends to hop from branch to branch rather than fly. Found in montane forest above 900m. Occurs in the Malay Peninsula and Sumatra; resident.

MALAYAN LAUGHINGTHRUSH *Garrulax peninsulae* 26cm

Newly recognised in 2006 as a species separate from northerly relatives, this laughingthrush is more skulking than other members of its genus. Overall olive-brown appearance with golden-olive wings and tail; wings with chestnut carpals, black coverts and golden flight feathers; dark chestnut crown and breast separated by grey face. Its various calls include a laughing hee hee hee that increases in volume, and three- to five-note whistles; seen in small parties near the ground. Found in montane forest, from about 900m upwards. Endemic to Malay Peninsula; resident

SILVER-EARED MESIA *Mesia argentauris* 17cm

Small flocks of mesias move through dense low vegetation, whistling and churring to each other. Black crown and face, with pale grey ear coverts; back and tail olive-grey; throat, breast and collar orange; primaries edged yellowish-orange with red bases, and tail-coverts red (male) or orange (female). Glimpsed in low flight can give overall olive appearance despite bright colours. Call a repeated whistle, *hello hello how-d'you-do*; in small parties in dense low bushes and ferns. Found in montane forest above 900m, especially forest edge, montane roadsides, tea estates, gardens. Occurs from the Himalayas to Sumatra; resident.

HIMALAYAN CUTIA *Cutia nipalensis* 18cm

Looking like a cross between a nuthatch and a shrike-babbler, the Cutia is one of the most sought-after birdwatching experiences. It is white below, with narrow black bars on sides of breast and flanks; back and rump is chestnut (male) or olive with black spots (female). Black mask with dusty blue crown; wings black, with blue bases and white tips to tail feathers; black tip of tail. Creeps slowly like a nuthatch along branches and tree trunks, usually solitary. Found in montane forest above 1,200m, keeping to larger trees. Occurs from the Himalayas to the Malay Peninsula; resident.

WHITE-BROWED SHRIKE-BABBLER *Pteruthius flaviscapis* 16cm

Fairly common in montane forest where it can be found in stunted elfin trees as well as tall forest, this species is bigger and stouter than other shrike babblers. Male is black above including crown and sides of face, with white eyebrow; rufous-buff patch on tips of secondaries, and faint white tips to primaries; below, the plumage is a very smooth-looking greyish-cream. Female is similar but with a washed-out appearance; olive-green wings with rufous patch on secondaries. Found in montane forest above 800m. Occurs from Pakistan to Borneo and Java; resident.

BLACK-EARED SHRIKE-BABBLER *Pteruthius melanotis* 12cm

This is the smaller and more brightly coloured of two shrike-babblers in the region. Like a little thick-billed warbler: brilliant yellow below, male more orange on its throat and upper breast, female whitish; ear coverts yellowish with a black outline. Crown and back olive-green, a faint slaty collar behind nape; wings dark with two bars (white in the male, buff in the female), and tail dark with white outer feathers. Found in montane forest above 1,100m altitude, taller trees as well as forest edge. Occurs from Burma to Indochina and the Malay Peninsula; resident.

BLUE-WINGED SIVA *Siva cyanouroptera* 16cm

Amongst the commoner birds at hill stations, Blue-winged Sivas are often overlooked because the blue in their plumage can be very hard to see. Slim, dark grey-brown bird with a rather long, square-ended tail. Pale eyebrow bordered above with black reminiscent of Mountain Fulvetta; usually pale blue-white eye; in good light, lilac-blue wash over flight-feathers of wing and tail. Call a thin but clear double whistle emphasising the first note, or multiple notes rising in pitch at the end; keeps to trees in small parties. Found in montane forest above 900m. Occurs from the Himalayas to Indochina and the Malay Peninsula; resident.

MOUNTAIN FULVETTA *Alcippe peracensis* 16cm

Fulvettas are amongst the commonest birds in the forest, but are not obvious. Greyish head with a black eyebrow beginning above eye continuing to nape; whitish to pale grey below, with buff tinge on abdomen and under tail-coverts; olive-brown back, wings and tail. Distinguished from Blue-winged Siva by olive-brown (not ashy) upperparts; no blue on wings and tail. Told from lowland Brown Fulvetta *A. brunneicauda* by black eyebrow. Moving through middle and lower storey in small parties, mixed feeding flocks with other species. Found in hill and montane forest above 300m to 1,500m. Occurs in Indochina and the Malay Peninsula; resident.

LONG-TAILED SIBIA *Heterophasia picaoides* 30cm

Small parties of sibias undulate from tree to tree, moving over open spaces and over the forest canopy in the same way as minivets. A slim, elongated bird, dark brownish-grey all over with a long graduated tail; small white patch at base of wing feathers, and white tips to tail feathers. Call a variety of sweet whistles, and a soft repeated *chip* contact call between members of a flock; feeding mainly in the tree-tops on small berries. Found in montane forest from about 900m upwards; conspicuous at hill stations. Occurs from the Himalayas to the Malay Peninsula; resident.

WHITE-BELLIED ERPORNIS *Erpornis zantholeuca* 12cm

The White-bellied Erpornis is a small, atypical babbler-like bird which is not closely related. Peaked crest and entire upperparts yellowish-green, face and underparts nearly white with yellow under tail-coverts. Lack of eyebrow, wingbars, rump and tail markings, and presence of crest, distinguish it from warblers. Singly and in mixed feeding flocks, from canopy to lower storey; call a high-pitched three-note churr. Found in forest, forest edge to about 1,000m (higher on isolated peaks). Occurs from the Himalayas to Taiwan, Sumatra, Borneo; resident.

ORIENTAL REED-WARBLER *Acrocephalus orientalis* 19cm

A rather big warbler, being about the size of a bulbul, which is plain brown with buff eyebrow, whitish-buff below becoming rusty buff on flanks; dusky colour between eye and bill forming lower border to the eyebrow. Call a series of loud harsh notes, a few croaks and a few high notes interspersed; comes briefly into view at top of reeds before plunging back into cover. Found in reed beds, grass and scrub, especially near old mining pools, rivers, canals. Breeds in Siberia, China, Mongolia, Japan and Korea, winters from south China and north-east India south to Indonesia and the Philippines; occasional migrant.

RUFESCENT PRINIA *Prinia rufescens* 12cm

This is a less common bird than the Yellow-bellied Prinia and somewhat patchily distributed. Similarly slim and active, smaller, plain rufous-brown above and cream below; head slightly greyer than rest of plumage especially in breeding season, with short whitish brow in front of eye; pale tips to tail feathers. Abdomen and under tail-coverts cream to buff, not yellow. Call like a tailor bird, *kedeek kedeek*. Found in forest edge, roadside and riverside scrub, overgrown agricultural land; short weak flight from clump to clump of scrub. Occurs from the Himalayas to the Malay Peninsula; resident.

YELLOW-BELLIED PRINIA *Prinia flaviventris* 14cm

This slim and active little bird is not always brightly coloured, and some individuals may be hard to tell from Rufescent Prinia. Slaty grey head, sometimes with a whitish brow in front of eye; olive upperparts, and cream breast grading into yellow or yellowish abdomen and under tail-coverts; narrow long tail. Distinguished from tailor birds by grey head, from Rufescent Prinia by yellowish underparts and olive (not rufous) tone to upperparts. Found in tall grass, roadside scrub, on agricultural land. Occurs from Pakistan east to Borneo and Java; resident.

MOUNTAIN TAILORBIRD *Phyllergates cucullatus* 12cm

A common bird of hill stations that must be separated from Yellow-bellied Prinia. Adult has chestnut crown, pale eyebrow, greyish nape and face grading into olive upperparts and whitish breast; entire abdomen, flanks and under tail-coverts yellow. Juvenile has olive-green crown. Distinguished from Yellow-bellied Prinia by chestnut crown of adult, olive (not grey) crown of juvenile. Distinguished from Chestnut-crowned Warbler by lack of black stripes on crown. Found in forest, forest edge, bamboo and roadside scrub in mountains above 1,000m. Occurs from the east Himalayas to Indochina, Borneo and Java; resident. The Philippines and Sulawesi; migrant.

COMMON TAILORBIRD *Orthotomus sutorius* 12cm

Their darting, low flight from thicket to thicket and continual change of perch make tailor birds hard to see well even when they are very close. This species olive above, creamy buff below including buff under tail-coverts, with rufous forehead and thighs, greyish face. Often shows some grey or even blackish feather-bases on throat but always less than Dark-necked Tailorbird; distinguished from that species also by thigh colour, lack of yellow on under tail-coverts, longer tail. Call a loud *kedeek kedeek kedeek*. Found in secondary growth, scrub, gardens, plantations. Occurs from India to Sumatra and Java; resident.

DARK-NECKED TAILORBIRD *Orthotomus atrogularis* 11cm

This is a bird of disturbed forest, rather than open country. Olive-green above, pale cream to dusky below with yellow under tail-coverts. Male has entire crown chestnut, and blackish throat and sides of neck; female like Common Tailorbird but has yellow beneath tail and lacks chestnut on thighs. Call a rolling *tttrrrrrit, tttrrrrrit*, repeated with variations; keeps to low growth along logging tracks, riverside growth. Found in forest edge and disturbed forest, often with bamboo or other invasive plant growth, also found in overgrown clearings. Occurs from north-east India to Sumatra and Borneo; resident.

ASHY TAILORBIRD *Orthotomus ruficeps* 12cm

The best chances of locating this fairly common grey and chestnut tailor bird are in mangroves. Adult has the entire head and face chestnut, nape and body grey becoming whitish on abdomen and under tail-coverts. Female paler below than male, and juvenile paler with whitish throat; juvenile distinguished from Rufous-tailed Tailorbird O. sericeus by lack of chestnut in tail. Calls include a fizz followed by a three-in-one trill, *pf-trt*, and a two-note trill *tree-dip*. Found low down in mangroves, growth behind mangroves, scrub, riverside vegetation. Occurs from southernmost Burma to Palawan and Java; resident.

ARCTIC WARBLER *Phylloscopus borealis* 12cm

Breeding widely across northern Old World, migrants in winter concentrate in south-east Asia. It is dusky olive above, with a pale yellowish eyebrow but no crown stripe, a single faint wingbar (a second one is sometimes visible further forwards); dull creamy white below with a brownish-olive flank. No pale rump patch. Call a brief rattle, rising at the end, or a single hoarse cheet. Found in forest edge, overgrown plantations, secondary growth, and migrates over forest from lowlands to 1,500m. Occurs across Eurasia to Alaska, south to Sulawesi in winter; migrant.

MOUNTAIN LEAF-WARBLER *Phylloscopus trivirgatus* 12cm

The only resident species of leaf-warbler in this area, this is a common bird in the mountains. Adult greenish above, yellowish below, with yellowish-green eyebrows and central crown stripe separated by black bands; no wingbar or rump patch, upper mandible dark, lower mandible yellowish. Juvenile has greener underparts; banding on head duller, less clear-cut. Feeds in canopy, amongst foliage and epiphytes, sometimes down to understorey; scolding call. Found in forest and forest edge, above 1,200m. Occurs from the Malay Peninsula to the Philippines and New Guinea; resident.

CHESTNUT-CROWNED WARBLER *Seicercus castaniceps* 10cm

There are two tiny yellow warblers that keep to thick montane scrub and must be distinguished by head colour. This species is greyish on the sides of face and breast, yellow on flanks, yellowish-olive above with yellow rump and two yellow wingbars; crown chestnut with a black line on each side of crown. The very similar Yellow-breasted Warbler *S. montis* has yellow breast and entirely chestnut head with black lines; both have white eye-ring. Found in hill and montane forest above 800m, keeping mainly to forest edge and understorey, and also to roadsides. Occurs from the Himalayas to Sumatra; resident.

ASIAN BROWN FLYCATCHER *Muscicapa dauurica* 14cm

A range of small greyish-brown flycatchers occurs in the region on migration, and the species are often difficult to tell apart. This species is greyish-brown above, unmarked whitish below with a grey wash on breast and flanks; whitish eye-ring. Bill dark with paler yellowish base to lower mandible; feet dark. Bill colour and paler breast and flanks distinguish from Dark-sided Flycatcher *M. sibirica*. Perches upright on low trees, twigs, sallying after insects. Found in forest edge, wooded gardens, mangroves. Occurs in north-east Asia, migrating to the whole of south and east Asia; various populations migrant here.

YELLOW-RUMPED FLYCATCHER *Ficedula zanthopygia* 13cm

This is the most colourful of the migrant flycatchers. Male black above with white eyebrow, white wing patch and yellow rump; yellow underparts grading into white under tail-coverts. Female olive-brown above with yellow rump; yellowish to buff underparts grading into white under tail-coverts, and one (sometimes two) narrow buff wingbars. Found wintering in lowland forest and tall secondary growth, to montane forest when on migration. Occurs from north-east Asia to the Malay Peninsula, Sumatra and Java; migrant.

RUFOUS-BROWED FLYCATCHER *Ficedula solitaris* 12cm

Rufous-browed Flycatchers allow a fairly close approach, like many montane birds. Olive-brown upperparts, becoming more rufous-brown on wings, rump and tail; contrasting white throat, centre of abdomen white, with buffish-brown wash on breast; bright rufous buff on forehead and sides of face, ring round eye. Juvenile has rufous streaks above and less contrast to forehead and eye-ring. Perches close to ground in forest understorey, nesting against steep banks. Call is a harsh *churr* or high-pitched whistle. Found in montane forest over 800m. Occurs from Burma and Indochina to Sumatra; resident.

LITTLE PIED FLYCATCHER *Ficedula westermanni* 11cm

This small chubby flycatcher has a particularly broad white eyebrow. Male black on crown and sides of face, back, wings and tail; broad white brow, white wingbar and white sides at base of tail; entirely whitish below. In the same habitat, distinguished from White-browed Flycatcher-shrike *Pteruthius flaviscapis* by long eyebrow, lack of olive in wing. Female small, pale greyish-brown with rump and tail more rufous; whitish below with grey tinge to breast; faint buff wingbar. Bill short, black. Found in montane forest and forest edge over 900m. Occurs from India east to Sulawesi and the Lesser Sundas; resident.

BLUE-AND-WHITE FLYCATCHER
Cyanoptila cyanomelana 17cm

This is an elegant flycatcher found as a migrant, so that its fine song is seldom heard here. Male deep blue on head, breast, back and wings; breast sharply divided from white abdomen; tail blue with white bases to outer feathers. Distinguished from White-tailed Flycatcher *Cyornis concreta* by smaller size and by sharp colour divide across breast. Female ashy grey-brown, paler brown beneath, off-white centre of throat. Found in lowland and lower montane forest, in middle storey. Occurs from north-east Asia, migrating through China to Sumatra, Borneo, Java and the Philippines; migrant.

LARGE NILTAVA *Niltava grandis* 21cm

This large, particularly dark flycatcher is common in montane forest. Male deep blue, often appearing black but in sunshine brilliant iridescent blue on crown, eyebrow and sides of neck. Female chocolate-brown, more rufous on wings and tail, with buff throat and bright blue patches on sides of neck. Call frequently heard, about four whistled notes ascending in scale. Found in montane forest and forest edge, lower and middle storey, above about 900m. Occurs from the Himalayas and south China to the Malay Peninsula and Sumatra; resident.

BLUE-THROATED FLYCATCHER *Cyornis rubeculoides* 15cm

This is one of the brightest, most iridescent of the blue and orange flycatchers in its genus. Male blue above, orange-rufous below, with shining blue eyebrow, rump, and upper tail-coverts, blue-black throat with a narrow orange triangle extending up chin nearly to bill. Female hard to identify, olive-brown above with faint rufescent tinge, especially on rump and upper tail-coverts; buffish-rufous throat grading into white abdomen. In middle and lower storey; song a series of high-pitched see-sawing notes. Found in lowland forest, forest edge. Occurs from India to Indochina and the Malay Peninsula; migrant.

GREY-HEADED CANARY-FLYCATCHER
Culicicapa ceylonensis 12cm

A common species in forest but one that is often overlooked, this flycatcher vaguely resembles a warbler because of its yellowish colouring. Head and upper breast ashy grey; back, wings and tail olive-green, lower breast and abdomen dull yellow. Rather erect posture and flycatching habit; call about six notes in three couplets, rising at the end; singly, in pairs and in mixed foraging flocks in middle and lower storey. Found in lowland and hill forest, tall secondary forest to about 1,000m. Occurs from Pakistan and India to Borneo and Java; resident.

115

GOLDEN-BELLIED GERYGONE *Gerygone sulphurea* 9cm

This tiny light yellow bird is hard to see but its call once learned, reveals it in a wide variety of habitats. Very small, olive-brown above, short rounded tail with pale subterminal mark on each feather; very bright yellow underparts, paler area between bill and eye. Call a series of hesitant scratchy whistles going up and down the scale, heard at all times of day; restless, quick-moving, often in canopy. Found in lowland and hill forest to about 800m, forest edge, plantations, mangroves, parks, gardens and roadside trees. Occurs from the Malay Peninsula to the Philippines and Sulawesi; resident.

BLACK-NAPED MONARCH *Hypothymis azurea* 16cm

This moderately sized flycatcher makes a small, inaccessible nest hanging from creepers. Male bright blue on head and breast, grading into duller blue on wings and tail, paler whitish abdomen; black patches on nape, round base of bill and band across breast. Female less intense blue, greyish-brown back, wings and tail, no black patches on head or breast; slender body, rather long, square-ended tail. Found in lowland and hill forest to about 1,100m, forest edge, overgrown plantations. Occurs from India to the Philippines and Java; resident.

RUFOUS-WINGED PHILENTOMA
Philentoma pyrhoptera 18cm

The least blue of the two flycatcher-like philentoma species has two poorly defined colour phases. Male either powder-blue on head, with chestnut wings and tail, and pale abdomen, or pale powder-blue all over, paler below; it is distinguished from the female Maroon-breasted Monarch by its lack of dark face. Female has greyish-brown head and breast, rufous wings and tail. Both sexes have reddish eyes, rather slim body. Singly, in pairs or mixed foraging flocks, rather sporadic. Found in middle and lower storey of lowland and hill forest to about 900m altitude. Occurs in the Malay Peninsula, Sumatra and Borneo; resident.

MAROON-BREASTED PHILENTOMA *Philentoma velata* 20cm

In this medium-sized but rather stocky species the male is overall deep matt blue with the sides of the face and throat black, merging into maroon breast patch. Female similar but duller blue all over, dark sides of face and throat; rump, abdomen and under tail-coverts bluish-grey. Both sexes have red-brown eyes. Call a musical single whistle and a harsh *churr*; also a long series of spaced, descending bell-like notes; in middle and lower storey. Found in lowland and hill forest up to 900m. Occurs in the Malay Peninsula, Sumatra, Borneo and Java; resident.

ASIAN PARADISE FLYCATCHER *Terpsiphone paradisi* 22-40cm

The Asian Paradise Flycatcher has very distinct colour phases and is widespread in lowland forest. Male has black head and shaggy peaked crest, bright blue skin round eye, blue bill. Body and very long ribbon tail either entirely white, with black wing feathers, or bright chestnut-brown body and tail, greyish face and breast, paler abdomen. Female like the chestnut-phase male, with shorter tail, shorter crest. Call a repeated harsh chack; in middle and lower storey, bamboo and dense undergrowth. Found in lowland and hill forest to about 1,100m, overgrown plantations, secondary forest. Occurs from Afghanistan to Borneo and Java; resident.

WHITE-THROATED FANTAIL *Rhipidura albicollis* 18cm

The skittish behaviour of fantails is as distinctive as their plumage, and is the reason for their Malay name of 'mad flycatcher'. Dark slaty grey above and below, with sharply defined short white eyebrow and white throat; long, wedge-shaped tail black with white tips to feathers. Singly or in pairs, or in mixed foraging flocks, flitting between perches while flicking and fanning tail; harsh churrs and short whistled song. Found in lower and middle storey of montane forest and forest edge above 700m. Occurs from the Himalayan foothills to China, Sumatra and Borneo; resident.

VELVET-FRONTED NUTHATCH *Sitta frontalis* 12cm

Once a good view of this creeping bird has been obtained, it is unmistakable. Entirely purplish-blue above, whitish to pearly grey below, with bright red bill and dark feet. A black velvety patch of feathers on forehead, extending back over eye in male. Small, often solitary but sometimes in mixed feeding flocks, creeping head first down living and dead tree trunks and branches, in upper and middle storey. Usually flies only to next tree. Does not call as such, but makes mouse-like squeaks. Found in lowland and hill forest and forest edge up to about 1,000m. Occurs from India to the Philippines and Java; resident.

BLUE NUTHATCH *Sitta azurea* 12cm

A soberly coloured nuthatch of the montane forest, appearing very dark in misty or gloomy weather. Creamy white throat and breast are sharply defined. Rest of plumage appears black, but in good light greyish-blue flight feathers are visible in wing, blue under tail-coverts, blue wash to abdomen and back. Bluish-white bill and skin round eye. Distinguished from Velvet-fronted Nuthatch by pied appearance, pale bill, dark legs. Usually solitary, creeping down larger branches and tree-trunks, in upper and middle storey. Found in montane forest above 900m. Occurs in Peninsular Malaysia, Sumatra, Java; resident.

119

YELLOW-THROATED FLOWERPECKER
Prionochilus maculatus 9cm

This tiny bird is amongst the commonest of flowerpeckers in undisturbed forest. Adult mainly olive-green, with short tail and short thick bill; underparts yellow in centre from breast to abdomen, whitish with dark olive streaks at sides; small yellow and orange patch on centre of crown, seldom visible. Juvenile is dark olive above, yellowish below, difficult to identify in field. Singly or in pairs, call not distinctive. Found in lowland and hill forest, forest edge, tall secondary vegetation, in middle and lower storey. Occurs in the Malay Peninsula, Sumatra and Borneo; resident.

CRIMSON-BREASTED FLOWERPECKER
Prionochilus percussu 10cm

This is a flowerpecker mainly of disturbed habitats, being found along logging tracks in the lowlands. Male greyish-blue above from crown to tail, yellow below with a red patch on centre of breast; red patch on top of crown and obscure white moustache streak. Female greyish-olive, paler and greyer below with yellowish centre down breast and abdomen; obscure orange patch on crown (hard to see) and whitish moustache streak. Found in lowland forest, forest edge in understorey. Occurs in the Malay Peninsula, Sumatra, Borneo, Java.

ORANGE-BELLIED FLOWERPECKER
Dicaeum trigonostigma 8cm

This flowerpecker has a wide altitudinal range, being found along the edges of most forest types. Male greyish-blue on head, upper back, wings and tail; pale grey throat grading into bright orange over breast and abdomen, orange lower back and rump. Female light olive with pale grey throat merging into pale yellow abdomen, yellow to orange rump. Bill is slightly more slender than other flowerpeckers. Singly or in pairs, low down in understorey and along forest edge; call a harsh clicking note. Found in lowland forest, secondary growth to at least 1,000m. Occurs from east India to the Philippines; resident.

SCARLET-BACKED FLOWERPECKER *Dicaeum cruentatum* 8cm

One of the commoner flowerpeckers in rural habitats, this is a familiar garden bird. Male tricoloured: brilliant scarlet from forehead to rump, black from sides of face down neck and wings to tail; whitish-grey from throat down centre of breast and abdomen. Female olive-grey with dark tail, bright red rump, pale whitish-grey below. Singly or in pairs, from low bushes to crowns of trees. Found in gardens, towns, cultivated areas, secondary vegetation from lowlands to about 1,300m. Occurs from India to south China, Sumatra and Borneo; resident.

BROWN-THROATED SUNBIRD *Anthreptes malacensis* 13cm

The Brown-throated Sunbird is one of the larger and commoner sunbirds to be found in gardens and open country. Male has iridescent green crown and back, purple rump and dark tail; sides of face olive. Throat dull brownish in centre with purple at sides of throat; rest of underparts bright yellow. Female has bright greenish-yellow underparts, no white markings in square-ended tail, no distinctive rump patch. Found in gardens, plantations, secondary vegetation, open scrub, mangroves. Occurs from Burma to Indochina, Philippines and Sulawesi; resident.

VAN HASSELT'S SUNBIRD *Leptocoma brasiliana* 10cm

Though it may at first appear black, good lighting shows this to be one of the most colourful sunbirds. Male largely iridescent, with glossy green crown, purple throat, crimson breast and abdomen merging into black under tail-coverts. Black back merging into glossy bluish-green wings, rump and tail. Female small, bright yellow, dark sides of face contrasting with pale yellow throat; tail dark and short with no white marks. Found in edges of plantations, secondary vegetation, gardens and mangroves. Occurs from north-east India to Sumatra and Borneo; resident.

OLIVE-BACKED SUNBIRD *Cinnyris jugularis* 12cm

This, the smallest of the sunbirds, is abundant and easily recognised in open country; also common in urban gardens and parks. Male olive-brown above, with glossy purplish-black throat and upper breast; remainder of underparts bright yellow. Female small, olive above and bright yellow below including yellow sides of face; tips of tail feathers dull white, best visible from below. Singly or in pairs, foraging on bushes and low trees, often briefly on exposed perches. Found in gardens, secondary vegetation, open country, mangroves. Occurs from south China to the Philippines and through to Australia; resident.

BLACK-THROATED SUNBIRD *Aethopyga saturata* 10–13cm

This is a common and beautiful sunbird found in the mountains. Male overall dark, with yellow underparts and rump; crown, nape and tail bright iridescent purplish-blue, with back and wings dark maroon, and rump yellow; tail with elongated central feathers. Face, throat and upper breast iridescent black, lower breast and abdomen are dull yellow with a faint streaking. Female olive-green with greyish throat, yellow colour below and yellow rump. Found in hill and montane forest above 500m, in understorey and forest edge. Occurs from the Himalayas and south China to the Malay Peninsula; resident.

123

TEMMINCK'S SUNBIRD *Aethopyga temminckii* 10–12cm

Very similar to Crimson Sunbird in appearance, this species is its equivalent in hill and montane forest. Male predominantly bright red, scarlet tail separated from back by blackish upper tail-coverts, narrow yellow rump. Iridescent violet crown and eyebrow often appear black. Female olive-green like Crimson Sunbird but head greyer, wings and tail tinged reddish, flanks tinged grey. Its chief food source are the flowers of flowering trees and shrubs. Found in hill and lower montane forest to about 1,500m, forest edge and along roadsides in montane areas. Occurs in the Malay Peninsula, Sumatra and Borneo; resident.

CRIMSON SUNBIRD *Aethopyga siparaja* 10–12cm

Recent taxonomic work has attempted to distinguish more carefully between the several bright red sunbirds in south-east Asia. Male of this species predominantly bright red with crimson back, yellow rump; dark green tail often appears black, with elongated central feathers. Iridescent green forehead; grey abdomen, and dark wings. Female is entirely dull olive-green, paler yellow below, with dull greenish under tail-coverts; it is distinguished from the female Brown-throated by duller underparts and small size. Found in forest edge, tall secondary growth, plantations. Occurs from India to the Philippines and Sulawesi; resident.

LITTLE SPIDERHUNTER *Arachnothera longirostra* 16cm

The Little Spiderhunter is the commonest of its group, and typical of disturbed forest where it feeds at banana inflorescences. Olive-green above, greyish on head with obscure moustache streak, throat whitish grading into olive breast and yellow abdomen and under tail-coverts. Little orange tufts occasionally visible at bend of wing, more often present in males than females. Call a loud *tchek*, uttered at intervals during rapid direct flight and sometimes monotonously when perched. Found in forest, logged and secondary forest to about 1,500m. Occurs from India to Java and the Philippines; resident.

LONG-BILLED SPIDERHUNTER *Arachnothera robusta* 21cm

The plumage is not very distinctive, but this species has the most impressive bill of all the spiderhunters. Large, with long, thick, black bill; dark olive-green patternless head, olive-green above, yellowish and faintly streaked below, becoming yellow on abdomen. Tail feathers have pale tips on underside. Usually feeds high in forest canopy, often perched on bare twigs; protects food sources aggressively; call a single harsh *chack*. Found in lowland, hill and lower montane forest up to 1,300m, commoner towards the upper end of this range. Occurs in the Malay Peninsula, Sumatra, Borneo and Java; resident.

SPECTACLED SPIDERHUNTER *Arachnothera flavigaster* 21cm

Bulkier than Yellow-eared Spiderhunter, this species often feeds lower down in the trees. Broad, complete yellow ring round eye, often linking up with small yellow ear patch; plumage olive above, pale olive-grey below without streaks; large size. Takes nectar from flowers of a variety of forest trees, often protecting food sources very aggressively. Call is a high-pitched *chit-chat*. Found in logged forest and secondary growth, coconut plantations and abandoned cultivation, mainly lowlands but occasionally over 1,000m. Occurs in the Malay Peninsula, Sumatra and Borneo; resident.

YELLOW-EARED SPIDERHUNTER
Arachnothera chrysogenys 17cm

Difficult to distinguish from the Spectacled Spiderhunter; special attention must be paid to its head pattern. Often incomplete yellow ring round eye, typically not linking up with yellow ear patch; plumage olive above, pale olive-grey below appearing faintly streaked, becoming yellowish on abdomen and beneath tail. Juvenile has eye-ring, but ear patch reduced. Feeds singly, mostly high up in forest canopy; call when perched a slow trill ending in a single long note. Found in lowland forest, forest edge, secondary vegetation, gardens up to 900m. Occurs in the Malay Peninsula, Sumatra, Borneo, Java; resident.

GREY-BREASTED SPIDERHUNTER *Arachnothera affinis* 17cm

A soberly plumaged bird, the Grey-breasted Spiderhunter is confined to the forest understorey and is scarcer than the Little Spiderhunter. Dull olive-green upperparts, whitish-grey to olive-grey underparts, the breast and crown narrowly streaked grey. Identified by cold colouring, streaking, and lack of other features; no yellow on underparts. Much less heavily streaked than Streaked Spiderhunter, feet not bright orange. Juvenile without streaks on breast. Found in forest, logged forest, occasionally reported from overgrown cultivation with bananas. Occurs in the Malay Peninsula, Sumatra, Borneo and Java; resident.

STREAKED SPIDERHUNTER *Arachnothera magna* 18cm

This finely plumaged spiderhunter, a montane resident which was once considered limited to the Main Range, has recently been found on the east coast mountains. Upperparts are olive-yellow, and underparts cream, streaked overall with black; its bright orange-yellow feet distinguish it from all other spiderhunters. Found singly, feeding mainly on nectar from banana flowers; flies in undulating pattern, call a rapid loud *chit chit chit* running together into a trill. Found in montane forest and forest edge above 900m. Occurs from the Himalayas to Indochina and the Malay Peninsula; resident.

JAVAN MUNIA *Lonchura leucogastroides* 11cm

A speciality of Singapore, the Javanese Munia has been introduced there and is spreading. Dark brown head, back, wings and tail, darker face and blackish-brown breast that is sharply cut off from white abdomen. During rice harvest forms flocks, but usually lives in pairs or small parties. Found in parks, gardens, secondary vegetation. Occurs from Java and Sumatra to Lombok; introduced resident. More widespread White-bellied Munia *L. leucogastra* is darker above with faint shaft streaks, dark breast breaks up into speckles on white abdomen, and tail has yellow tinge.

SCALY-BREASTED MUNIA *Lonchura punctulata* 11cm

One of the commoner munias in gardens and towns, this bird can often be seen on overgrown roadside verges. Adult plain brown above, on wings, rump and tail, face and throat darker, more rufous; underparts dirty white with each feather margined and centred dark brown, producing a scaly appearance; centre of abdomen white. Juvenile plain fawn-brown, bill black above and light below. Found in gardens, roadsides, rice fields, scrub and secondary vegetation. Occurs from India to Taiwan and the Philippines, introduced to various countries; resident.

BLACK-HEADED MUNIA *Lonchura atricapilla* 11cm

Perhaps the commonest of all munias in rural areas. The black head of this species is the best feature to look for when the bird is perched or during the whirring bee-like flight of flocks in grassland. Black head and upper breast, sharply defined from rich brown body, wings and tail. Juvenile overall warm fawn-brown, tends to be darker on head. Bill thick, conical, bluish-grey. Forages on grass seed-heads and rice, forming big flocks. Found in gardens, scrub, secondary growth and rice fields, plantation edges and swamps. Occurs from India to the Philippines and Sulawesi, introduced elsewhere; resident.

WHITE-HEADED MUNIA *Lonchura maja* 11cm

Easily identified as a reverse of the Chestnut Munia in colouring, this species is not quite so common. Adult with head white or whitish, merging gradually into chestnut-brown body, darker below merging into black under tail-coverts. Bill thick, conical, bluish-grey. Juvenile warm fawn all over, head slightly paler than body and best identified by association with adults. Forages on grass seed-heads, often mixed with other munias. Found in grassland, rice fields, swamp, open secondary growth. Occurs in the Malay Peninsula, Sumatra and Java; resident.

JAVA SPARROW *Padda oryzivora* 16cm

Introduced at various times and places, this bird seems to make headway for a few years as a colonist before diminishing again. Adult grey with black head and pure white cheeks, black tail and white under tail-coverts, thick conical pink or red bill. Juvenile browner, with buff or brown breast. In small parties or sometimes singly, feeding on rice and grass seed-heads, or in low bushes and scrub. Found near towns, villages and sometimes rice fields, occurring sporadically. Occurs in Java and Bali, widely introduced elsewhere; resident.

EURASIAN TREE SPARROW *Passer montanus* 14cm

Amongst the commonest of all birds in the region, Eurasian Tree Sparrows are excellent colonists and make their way even to remote clearings. Crown chestnut; throat, sides of face and ear patch black, contrasting with cold buff underparts; back and wings brown mottled with black; wings with two pale bars. Commonly feeding on ground, active even at night in bright cities. Found in all open habitats from scrub and coasts to gardens, cultivation, towns, reaching clearings and villages in mountains. Occurs throughout Europe and Asia to Sumatra and Java, introduced to Borneo, North America and Australia; resident.

BAYA WEAVER *Ploceus philippinus* 15cm

Patchily distributed in the rural lowlands, small colonies of these birds build their nests in isolated low trees and in coconut palms. Breeding male has bright yellow crown, dark brown mask covering sides of face, underparts plain buff, upperparts mottled dark brown and buff. Female and non-breeding male similar but entire head tawny brown, no face mask, eyebrow rufous. Both sexes have thick conical bill. Feeding in grass, rice, and on ground. Found in scrub, edges of cultivation, rice fields, scattered trees. Occurs from Pakistan to Sumatra and Java; resident.

ORIENTAL WHITE-EYE *Zosterops palpebrosus* 11cm

This is a popular cage-bird found in disturbed vegetation, overlapping with similar Everett's White-eye *Z. everetti* in the mountains. Bright olive-green above, bright yellow on throat merging into grey abdomen and under tail-coverts; wings and tail darker. Bright white feathering round eye, set off by darker sides of face; forehead often with traces of yellow. Everett's White-eye has darker upperparts, no yellow on forehead, darker grey abdomen. Found in mangroves, forest edge, logging tracks from coast up to about 1,800m. Occurs from Afghanistan to Borneo and Java; resident; in Singapore individuals in the wild are considered as escaped from captivity.

131

ASIAN GLOSSY STARLING *Aplonis panayensis* 20cm

Noisy roosting flocks that include black adults and speckled juveniles can often be found on high-tension wires. Adult entirely black with an oily green gloss all over; eyes bright red. Juvenile (right photo) dark greenish-grey above, somewhat glossy, and dirty buff with grey streaks below. Slim, small-headed; flight direct, rapid wingbeats showing small triangular wings and short tail; in small flocks, often giving noisy piping notes. Feeds mainly on soft fruits and rarely visits the ground. Found in secondary vegetation, gardens, towns, open country including coastal areas and small islands. Occurs from India to the Philippines and Sulawesi; resident.

COMMON MYNA *Acridotheres tristis* 25cm

Currently the most widespread myna; its rich brown plumage and yellow eye-skin are key characteristics. Adult has black head, grading into plumbeous brown body, white under tail-coverts; yellow bill, legs and skin round eye; bald yellow-headed individuals are often seen. In flight, conspicuous white wing patches, white tip to tail, undersides of wings white. Juvenile duller, greyer brown, head and breast brown; no bare skin round eye. Call of mixed harsh and melodious whistles; forages mainly on short grass. Found in towns, villages, gardens and all open habitats. Occurs from Afghanistan to the Malay Peninsula, and widely introduced; resident.

JUNGLE MYNA *Acridotheres fuscus* 23cm

Apparently losing ground against other mynas, this species is becoming progressively scarcer and more difficult to identify as other species are introduced. Slightly smaller than Common Myna, greyish- (not plumbeous) brown with dark head and short crest over bill; wing patch, tail tip, abdomen and under tail-coverts white. Eye yellow without bare surrounding skin, bill yellowish, with blue base to lower mandible characteristic but hard to see. Juvenile duller, browner with more extensive whitish below. Forages on ground and in low trees. Found in gardens, plantations, rice fields and secondary scrub. Occurs from India to the Malay Peninsula; resident.

JAVAN MYNA *Acridotheres javanicus* 25cm

Two introduced species have greatly complicated the identification of foraging roadside mynas. Javan Myna introduced to Singapore has now spread north beyond Kuala Lumpur, is ash-grey, darker than Jungle Myna and lacks brown tones; short crest. White-vented Myna *A. grandis*, introduced to Kuala Lumpur, is coal-black, with long shaggy crest hanging over the bill. Both have white wing patch, tail tip and under tail-coverts. Hybridisation may now be occurring. Found in towns, roadside verges, foraging mainly on short grass, and in secondary scrub, cultivation and shorelines. Occurs from Bangladesh to Indochina, Java and Sulawesi; resident and spreading.

COMMON HILL-MYNA *Gracula religiosa* 30cm

Still fairly common in lowland forest and wooded rural areas, hill-mynas' loud calls are unmistakable. Heavy; plumage entirely black with purplish gloss, white patches at base of primaries; bill orange to yellow, feet yellow, and flaps of bare yellow skin on face and nape. Feeds mainly on fruit in canopy of trees, call typically a ringing *tiong*, the wings thrumming in flight. Found in tall lowland forest, forest edge, isolated trees in rural areas, often perched on dead branches and nesting in tree holes. Occurs from India to Borneo, Java and Palawan, introduced in some other countries; resident.

SULTAN TIT *Melanochlora sultanea* 18cm

Seldom seen but well worth the effort; the best views of this bulky tit are usually along the edges of logging tracks in hilly forest. Male intense black with bright yellow shaggy crest, yellow lower breast and abdomen. Female less intense brownish-black, upper breast olive-brown not black. Moving rapidly, usually alone, high up in forest canopy; call an attractive sliding whistle, rapidly repeated. Found in lowland and hill forest, up into lower montane forest. Occurs from the Himalayas to the Malay Peninsula, perhaps Sumatra; resident.

BLACK-NAPED ORIOLE *Oriolus chinensis* 26cm

Population due partly to escaped cage-birds and partly to recent natural extensions of range. The male is brilliant yellow with black mask through eyes meeting at nape, black wings, black tail with yellow tips; bill pinkish-orange. Females duller, the back and underparts yellowish-green to dull olive. Juveniles duller still, olive above. Call is a melodious four-note whistle, *What the devil!* with much individual variation. Found in mangrove edge, scrub, secondary growth and gardens, into city centres. Occurs from India to the Philippines; resident in coastal plains, supplemented by migrants from islands to far inland.

BLACK-AND-CRIMSON ORIOLE *Oriolus cruentus* 22cm

Often seen in silhouette or in poor misty conditions, this apparently all-black bird is worth a second and third look. Male intense black with big rounded crimson patch on the breast, and crimson flash on primary coverts. Female entirely black with the breast and abdomen greyish. Both sexes have blue feet and pale blue bill. Call a simple wheeze or plaintive mew, unlike bell-like notes of most other orioles. Found in tall montane forest above about 700m, keeping to dark middle storey. Occurs in the Malay Peninsula, Sumatra, Borneo and Java; resident.

BLACK DRONGO *Dicrurus macrocercus* 27cm

There are two open-country drongos of similar size, this being the darker; it has greatly expanded its range with increasing cultivation of the west coast plains. A slim, all-black drongo with a rather small bill, the tail deeply forked and the outer feathers slightly turned upwards at tips. As with other drongos, juvenile has inconspicuous whitish scales on abdomen and under wings. Perches conspicuously on wires, fences, bare trees, occasionally on domestic animals. Found in cultivated land, rice fields, scrub, secondary growth especially near coast. Occurs from the Middle East to China and Java; migrant southwards along west coast. Similar Ashy Drongo *D. leucophaeus* is usually paler grey, can look black but has red eyes; tail tips not turned upwards.

CROW-BILLED DRONGO *Dicrurus annectans* 28cm

One of the less common drongos, this is most likely to be seen in the middle and lower storey of hill forest. Similar to Black Drongo but heavier, with stouter bill; tail not so long and less deeply forked, with outer tips strongly turned upwards. Juvenile has white spotting on underparts. Distinguished from the Bronzed Drongo *D. aeneus* by its heavier bill and more upturned tail tips. Its calls include some twanging notes descending in pitch, also varied rasps and bell-like notes. Found in hill forest up to 1,000m during passage, but lower down in mangroves, secondary growth and plantations during the wintering period. Occurs from the Himalayas to Borneo and Java; passage migrant and non-breeding visitor.

LESSER RACKET-TAILED DRONGO *Dicrurus remifer* 25–65cm

The two drongos with racquet-tipped tail feathers are actually fairly simple to distinguish. This species is small, and a heavy tuft of bristles at the base of the bill gives a large-billed and beetle-browed appearance; tail square, not forked, with two outer tail feathers having very elongated bare shafts (up to 40cm) and a long oval-shaped racket that is flat or slightly twisted. Also separated by altitude, this species from about 900m upwards, in middle and upper storey, singly or in pairs, giving brief grating calls. Found in montane forest. Occurs from India and Nepal east to Java; resident.

GREATER RACKET-TAILED DRONGO
Dicrurus paradiseus 32–57cm

The most conspicuous and well-known drongo is also the biggest in the region and the most spectacular. All-black plumage, large size and somewhat rounded crown; tail forked not square, the two outer feathers having elongated bare shafts (up to 25cm) and large rounded and twisted racquets. Many birds have one or both rackets broken off, but traces of elongated broken shafts are usually visible, and large size, profile and forked tail without any upturning at tips are then distinctive. Gives a great variety of harsh and bubbling, bell-like notes, calls frequently. Found in lowland forest, plantations, secondary growth. Occurs from India to Borneo and Java; resident.

CRESTED JAY *Platylophus galericulatus* 32cm

Though it can move rapidly and inconspicuously through the forest, the Crested Jay's calls, once learned, reveal it to be reasonably common. Adult all dark, nearly black, with small white patch on sides of neck, white marks above and below eye, and a very long spatulate crest standing straight up over crown. Juvenile browner above, with rufous tips to wing-coverts, centre of belly pale to barred, and crest shorter. Call a long, harsh chatter like a shrike. Found in lowland forest to about 800m, mainly middle storey. Occurs in the Malay Peninsula, Sumatra, Borneo and Java; resident.

COMMON GREEN MAGPIE *Cissa chinensis* 38cm

Remarkably hard to see despite its colouring, the Common Green Magpie moves quietly through the middle and lower storey searching for insects and fruits. Lime-green plumage with bright red bill and a black mask through eye to nape; wings chestnut tipped with black and white on secondaries; tail long, tapering, the feathers tipped black and white. Call a series of harsh chatters followed by a whistle, or the chatters or whistle separately. Found in montane forest, especially lower storey, about 900–1,800m. Occurs from the Himalayas to Borneo; resident.

HOUSE CROW *Corvus splendens* 42cm

First introduced in vain as a pest-control agent, House Crows have spread now to many towns and agricultural areas in the west coast lowlands. Dark plumage, with traces of a grey band behind nape and across breast. The grey area is virtually absent in juveniles, strongly marked in some adults. Distinguished from Large-billed Crow by size, smaller bill, grey in plumage. Feeds on rubbish, road kills, oil-palm fruits and other opportunistic sources. Forms noisy communal roosts in towns. Found in mangroves, coastal and lowland towns, along roads, plantations, gardens throughout west coast. Occurs from Iran to China, introduced and still spreading as far as Australia.

SOUTHERN JUNGLE CROW *Corvus macrorhynchos* 50cm

With the advent of House Crows, this species has become known as a rural bird, but it can still be found occasionally in cities as well. Big, entirely glossy black; heavy bill with bristly base and high arched profile; tail rather long, wedge-shaped. Call deeper than other crows; not forming big flocks. Size, bill shape and deep voice distinguish it from both House Crow and the all-black Slender-billed Crow *C. enca*. Found in open country, plantations, secondary growth, mangroves, occasionally up to 1,800m at hill stations. Occurs from southern Thailand to the Sundas; resident.

FURTHER READING

The following books and other publications should be of interest to those wishing to learn more about birds, birdwatching and other wildlife in the region. The key bird field guide is Robson's *A Field Guide to the Birds of South-East Asia*.

Bransbury, J. *A Birdwatcher's Guide to Malaysia*. Waymark Publishing, 1993

Briffett, C. *A Guide to the Common Birds of Singapore*. Singapore Science Centre, Singapore, 1986

Das, I. *A Field Guide to the Reptiles of South-East Asia*. New Holland Publishers, 2010.

Francis, C.M. *A Field Guide to the Mammals of South-East Asia*. New Holland Publishers, 2008.

Hails, C.J. and Jarvis, F. *Birds of Singapore*. Times Editions, Singapore, 1987

Howard, R. and Moore, A. *A Complete Checklist of the Birds of the World*, 2nd edition. Oxford University Press, Oxford 1991

Lim, K.S. *Vanishing Birds of Singapore*. Nature Society (Singapore), Singapore, 1992

Myers, S. *A Field Guide to the Birds of Borneo*. New Holland Publishers, 2010.

Robson, C. *A Field Guide to the Birds of South-East Asia*. New Holland Publishers, 2011.

Robson, C. *A Field Guide to the Birds of Thailand*. New Holland Publishers, 2004.

Strange, M. and Jeyarajasingam, A. *Birds, a Photographic Guide to the Birds of Peninsular Malaysia and Singapore*. Sun Tree Publishing, Singapore, 1993

Tweedie, M.W.F. *Common Birds of the Malay Peninsula*. Longmans, Singapore, 1970

White, T. *A Field Guide to the Bird Songs of South-east Asia*. National Sound Archives, London, 1984

INDEX

Aceros corrugatus 64
   undulatus 64
Acridotheres fuscus 133
   grandis 133
   javanicus 133
   tristis 132
Acrocephalus
   orientalis 107
Actitis hypoleucos 37
Aegithina tiphia 88
Aethopyga
   saturata 123
   temminckii 124
   siparaja 124
Alcedo atthis 59
   meninting 59
Alcippe peracensis 105
Alophoixus
   ochraceus 86
Amaurornis
   phoenicurus 29
Anorrhinus galeritus 63
Anthracoceros
   albirostris 65
   malayanus 65
Anthreptes malacensis 122
Anthus rufulus 93
Aplonis panayensis 132
Apus affinis 57
Aquila clanga 24
Arachnothera
   affinis 127
   chrysogenys 126
   flavigaster 126
   longirostra 125
   robusta 125
Ardea alba 16
   cinerea, 15
   purpurea, 15
Ardeola bacchus 18
Arenaria interpres 38
Argus, Great 29
Argusianus argus 29

Babbler, Chestnut-winged 101
   Golden 100
   Grey-headed 101
   Grey-throated 100
   Horsfield's 97
   Rufous-crowned 98
   Sooty-capped 97
Barbet,
   Black-browed 69
   Coppersmith 69
   Fire-tufted 67
   Gold-whiskered 68
   Lineated 67
   Red-throated 68
Barn-owl,
   Common 53
Bee-eater,
   Blue-tailed 62
   Blue-throated 62
Berenicornis comatus 63
Bittern, Cinnamon 20
   Yellow 19
Blythipicus
   rubiginosus 73
Boobook, Brown 55
Booby, Brown 14
Broadbill,
   Black-and-red 74
   Black-and-yellow 74
   Green 76
   Long-tailed 75
   Silver-breasted 75
Bubo sumatranus 54
Bubulcus
   coromandus 17
Buceros bicornis 66
   rhinoceros 66
Bulbul, Ashy 88
   Black-and-white 80
   Black-crested 81
   Black-headed 80
   Cream-vented 84
   Hairy-backed 86
   Mountain 87
   Ochraceous 86
   Olive-winged 84
   Puff-backed 82
   Red-eyed 85
   Red-whiskered 82
   Scaly-breasted 81
   Spectacled 85
   Straw-headed 79
   Streaked 87
   Stripe-throated 83
   Yellow-vented 83
Butorides striata 18

Cacomantis sonnerati 49
Calidris ferruginea 39
Caloenas nicobarica 44
Calyptomena viridis 76
Canary-flycatcher,
   Grey-headed 115
Caprimulgus
   macrurus 56
Centropus
   bengalensis 52
   sinensis 52
Chalcophaps indica 43
Charadrius dubius 33
   mongolus 34
Chloropsis
   cochinchinensis 90
   cyanopogon 89
   hardwickii 90
   sonnerati 89
Chrysophlegma
   flavinucha 72
   miniaceus 71
Ciconia stormi 21
Cinnyris jugularis 123
Cissa chinensis 138
Collared-dove, Red 41
Collocalia esculenta 57
Columba livia 41
Copsychus
   malabaricus 94
   pyrropyga 95
   saularis 94
Coracina javensis 77
Corvus
   macrorhynchos 139
   splendens 139
Coucal, Greater 52
   Lesser 52
Crow, House 139
   Southern Jungle 139
Cuckoo,
   Banded Bay 49
Cuckoo-dove,
   Little 42
Cuckooshrike,
   Javan 77
Culicicapa
   ceylonensis 115
Cutia,
   Himalayan 103
Cutia nipalensis 103
Cyanoptila
   cyanomelana 114
Cymbirhynchus
   macrorhynchus 74
Cyornis
   rubeculoides 115

Dendrocopos
   moluccensis 70

Dendrocygna
  javanica 21
Dicaeum
  cruentatum 121
  trigonostigma 121
Dicrurus annectans 136
  leucophaeus 136
  macrocercus 136
  paradiseus 137
  remifer 137
Dinopium javanense 72
Dove, Emerald 43
  Spotted 42
  Zebra 43
Dowitcher, Asian 39
Drongo, Ashy 136
  Black 136
  Crow-billed 136
  Greater Racket-tailed 137
  Lesser Racket-tailed 137
Ducula aenea 46
  badia 47
  bicolor 47

Eagle, Greater Spotted 24
Eagle-owl, Barred 54
Egret,
  Eastern Cattle 17
  Great 16
  Intermediate 16
  Little 17
Egretta garzetta 17
Elanus caeruleus 22
Enicurus ruficapillus 95
Erpornis,
  White-bellied 106
Erpornis zantholeuca 106
Eudynamys scolopaceus 49
Eurylaimus ochromalus 74

Fairy-bluebird,
  Asian 91
Fantail, White-throated 118
Ficedula solitaris 113
  zanthopygia 112
Finfoot, Masked 31
Fireback, Crested 27
  Crestless 27
Fish Owl, Buffy 54
Flameback, Common 72

Flowerpecker,
  Crimson-breasted 120
  Orange-bellied 121
  Scarlet-backed 121
  Yellow-throated 120
Flycatcher,
  Asian Brown 112
  Blue-and-white 114
  Blue-throated, 115
  Little Pied, 113
  Rufous-browed 113
  Yellow-rumped 112
Forktail,
  Chestnut-naped 95
Fruit-dove, Jambu 46
Fulvetta,
  Mountain 105

Gallicrex cinerea 30
Gallinago stenura 38
Gallinula chloropus 30
Gallus gallus 26
Garrulax peninsulae 102
Geopelia striata 43
Gerygone,
  Golden-bellied 116
Gerygone sulphurea 116
Godwit,
  Black-tailed 34
Gracula religiosa 134
Grebe, Little 14
Green-pigeon,
  Little 44
  Pink-necked 45
  Thick-billed 45
Greenshank, Common 36

Halcyon pileata 61
  smyrnensis 60
Haliaeetus leucogaster 23
Haliastur indus 23
Hanging-parrot,
  Blue-crowned 48
Harpactes diardii 58
  duvaucelii 58
Hawk-eagle,
  Blyth's 25
  Changeable 25
Heliopais personata 31
Hemixos flavala 88
Heron, Grey 15
  Little 18

  Purple 15
Heterophasia picaoides 106
Hill-myna,
  Common 134
Himantopus himantopus 32
Hirundo tahitica 73
Hornbill, Black 65
  Bushy-crested 63
  Great 66
  Oriental Pied 65
  Rhinoceros 66
  White-crowned 63
  Wreathed 64
  Wrinkled 64
Hydrophasianus chirurgus 32
Hypothymis azurea 116

Imperial-pigeon,
  Green 46
  Mountain 47
  Pied 47
Iora, Common 88
Irena puella 91
Ixobrychus cinnamomeus 20
  sinensis 19
Ixos malaccensis 87
  mcclellandii 87

Jacana,
  Pheasant-tailed 32
Jay, Crested 138
Junglefowl, Red 26

Kenopia striata 99
Ketupa ketupu 54
Kingfisher,
  Black-capped 61
  Blue-eared 59
  Collared 61
  Common 59
  Stork-billed 60
  White-throated 60
Kite, Black-shouldered 22
  Brahminy 23
Koel, Asian 49

Lalage nigra 77
Lanius schach 92
  tigrinus, 91
Laughingthrush,
  Spectacled 102
  Malayan 102

142

Leafbird,
  Blue-winged 90
  Greater Green 89
  Lesser Green 89
  Orange-bellied 90
Leaf-warbler,
  Mountain 111
Leptocoma brasiliana 122
Limnodromus semipalmatus 39
Limosa limosa 34
*Lonchura
  atricapilla* 129
  maja 129
  punctulata 128
*Lophura
  erythrophthalma* 27
  ignita 27
Loriculus galgulus 48

Macronous gularis 101
  ruficeps 42
Magpie, Common
  Green 138
Magpie-robin,
  Oriental 94
Malacocincla sepiarium 97
Malacopteron affine 97
  magnum 98
Malkoha, Chestnut-
  bellied 50
  Chestnut-breasted 51
  Green-billed 50
  Raffles's 51
*Megalaima
  chrysopogon* 68
  haemacephala 69
  lineata 67
  mystacophanos 68
  oorti 69
Melanochlora sultanea 134
Merops philippinus 62
  viridis 62
Mesia argentauris 103
Mesia,
  Silver-eared 103
Mesophoyx intermedia 16
Minivet, Ashy 78
  Grey-chinned 79
Monarch,
  Black-naped 116
Moorhen,
  Common 30

Motacilla tschutschensis 93
Munia,
  Black-headed 129
  Javan 128
  Scaly-breasted 128
  White-headed 129
Muscicapa dauurica 112
Mycteria cinerea 20
Myna, Common 132
  Javan 133
  Jungle 133
  White-vented 133

Napothera brevicaudata 99
Night-heron,
  Black-crowned 19
Nightjar,
  Large-tailed 56
Niltava, Large 114
Niltava grandis 114
Ninox scutulata 55
Nisaetus alboniger 25
  limnaeetus 25
Numenius phaeopus 34
Nuthatch, Blue 119
  Velvet-fronted 119
Nycticorax nycticorax 19

Oriole, Black-and-
  crimson 135
  Black-naped 135
Oriolus chinensis 135
  cruentus 135
*Orthotomus
  atrogularis* 109
  ruficeps 110
  sutorius 109
Osprey 22
Otus lettia 53

Padda oryzivora 130
Pandion haliaetus 22
Paradise-flycatcher,
  Asian 118
Parakeet,
  Long-tailed 48
Partridge, Crested 26
Passer montanus 130
Peacock-pheasant,
  Malayan, 28
  Mountain, 28
Pelargopsis capensis 60
Pericrocotus divaricatus 78

  solaris 79
Philentoma, Maroon-
  breasted 117
  Rufous-winged 117
Philentoma pyrhoptera 117
  velata 117
Phyllergates cucullatus 108
Phylloscopus borealis 110
  trivirgatus 111
Piculet, Speckled 70
Picumnus innominatus 70
Picus puniceus 71
Pigeon, Nicobar 44
  Rock 41
Pipit, Paddyfield 93
Pitta, Banded 76
  Blue-winged 77
Pitta guajana 76
  moluccensis 77
Platylophus galericulatus 138
Ploceus philippinus 131
Plover,
  Little Ringed 33
  Pacific Golden 33
Pluvialis fulva 33
Polyplectron inopinatum 28
  malacense 28
Pomatorhinus montanus 98
Pond-heron,
  Chinese 18
Porphyrio porphyrio 31
Prinia, Rufescent 107
  Yellow-bellied 108
Prinia flaviventris 108
  rufescens 107
Prionochilus maculatus 120
  percussus 120
Psarisomus dalhousiae 75
Psilopogon pyrolophus 67
Psittacula longicauda 48
Pteruthius flaviscapis 104, 113
  melanotis 104
Ptilinopus jambu 46
Pycnonotus atriceps 80
  brunneus 85
  eutilotus 82

143

erythrophthalmos 85
finlaysoni 83
flaviventris 81
goiavier 83
jocosus 82
melanoleucos 80
plumosus 84
simplex 84
squamatus 81
zeylanicus 79

Redshank,
  Common 35
Reed-warbler, Oriental 107
Rhinocichla mitrata 102
Rhinortha chlorophaeus 51
Rhipidura albicollis 118
Rhopodytes sumatranus 50
  tristis 50
Rollulus rouloul 26

Sandpiper,
  Common 37
  Curlew 39
  Terek 37
  Wood 36
Sand-plover, Lesser 34
Saxicola maurus 96
Scimitar-babbler, Chestnut-backed 98
Scops-owl, Collared 53
Sea-eagle, White-bellied 23
Seicercus castaniceps 111
Serilophus lunatus 75
Serpent-eagle, Crested 24
Shama,
  Rufous-tailed 95
  White-rumped 94
Shrike, Brown 92
  Tiger 91
Shrike-babbler,
  Black-eared 104
  White-browed 104
Sibia,
  Long-tailed 106
Sitta azurea 119
  frontalis 119

Siva,
  Blue-winged 105
Siva cyanouroptera 105
Snipe, Pintail 38
Sparrow, Java 130
Spiderhunter,
  Grey-breasted 127
  Little 125
  Long-billed 125
  Spectacled 126
  Streaked 127
  Yellow-eared 126
Spilornis cheela 24
Stachyridopsis chrysaea 100
  erythroptera 101
  nigriceps 100
  poliocephala 101
Starling,
  Asian Glossy 132
Sternula albifrons 40
Stilt, Black-winged 32
Stonechat, Eastern 96
Stork, Milky 20
  Storm's 21
Streptopelia chinensis 42
  tranquebarica 41
Strix leptogrammica 56
  seloputo 55
Sula leucogaster 14
Sunbird,
  Black-throated 123
  Brown-throated 122
  Crimson 124
  Olive-backed 123
  Temminck's 124
  Van Hasselt's 122
Swallow, Pacific 73
Swamphen, Purple 31
Swift, House 57
Swiftlet, Glossy 57

Tachybaptus ruficollis 14
Tailorbird, Ashy 110
  Black-necked 109
  Long-tailed 109
  Mountain 108
Tern,
  Great Crested 40
  Little 40
Terpsiphone paradisi 118
Thalasseus bergii 40
Thrush, Chestnut-capped 96

Tit, Sultan 134
Tit-babbler,
  Pin-striped 101
Todirhamphus chloris 61
Tree-sparrow, Eurasian 130
Treron curvirostra 45
  olax 44
  vernans 45
Tricholestes criniger 86
Triller, Pied 77
Tringa glareola 36
  nebularia 36
  totanus 35
Trogon, Diard's 58
  Scarlet-rumped 58
Turnstone, Ruddy 38
Tyto alba 53

Wagtail,
  Eastern Yellow 93
Warbler, Arctic 110
  Chestnut-crowned 111
Waterhen,
  White-breasted 29
Watercock 30
Weaver, Baya 131
Whimbrel 34
Whistling-duck, Lesser 21
White-eye,
  Oriental 131
Woodpecker,
  Banded 71
  Crimson-winged 71
  Maroon 73
  Sunda Pygmy 70
Wood-owl,
  Brown 56
  Spotted 55
Wren-babbler,
  Streaked 99
  Striped 99

Xenus cinereus 37

Yellownape,
  Greater 72

Zanclostomus curvirostris 51
Zoothera interpres 96
Zosterops palpebrosus 131